The Good Health Garden

Growing and Using Healing Foods

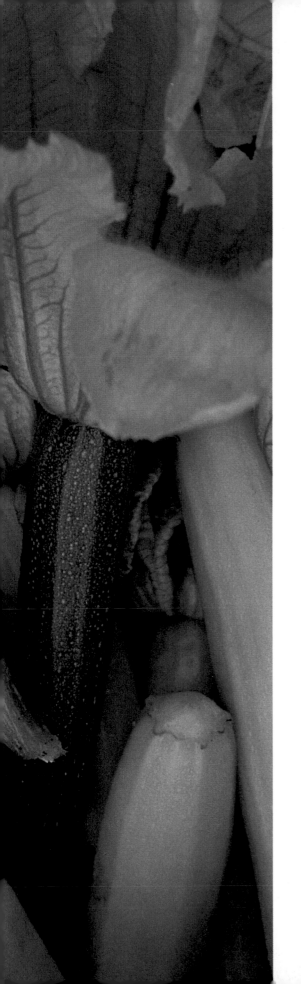

The Good Health Garden

Growing and Using Healing Foods

Anne McIntyre

THE READER'S DIGEST ASSOCIATION INC.
Pleasantville, New York • Montreal

A READER'S DIGEST BOOK

Designed and produced by Breslich & Foss Limited

Project editor: Laura Wilson
Copy editors: Jonathan Hilton, Caroline Taylor
Designer: Margaret Sadler
All photography by Juliette Wade except for p. 105 (top right), The
Garden Picture Library; p. 97 and p. 106 (top right) A-Z Botanical
Collection Ltd.
Illustrators: Madeleine David, Amanda Patton,
Polly Raynes

Library of Congress Cataloging in Publication Data

McIntyre, Anne.
    The good health garden : growing and using healing foods :
herbs. fruits, vegetables / Anne McIntyre
        p. cm.
    Includes bibliographical references and index.
    ISBN 0-7621-0016-8
    1. Vegetables—Therapeutic use. 2. Fruit—Theraputic use.
3. Herbs—Therapeutic use. 4. Organic gardening. I. Title.
    RM236.M395 1998
    615'.321—dc21   97-31949

Printed in Belgium

Contents

Introduction

'The love of gardening is a seed that once sown never dies.'

Gertrude Jekyll

Growing fruits, vegetables, and herbs in the garden to provide foods and medicines throughout the year is something that our ancestors did for thousands of years. The first gardeners are thought to have lived around ten thousand years ago, in the Middle East, S.E. Asia, and Central America, and we know that by Roman times many vegetables which are common today, such as turnips, cabbages, radishes, garlic, lettuce, chicory, gourds, and asparagus were cultivated in European gardens. It is only really over the last century that there has been a shift away from growing food for the kitchen in individual vegetable plots to mass production of food for our increasing urban populations. Now, there is growing disenchantment with the use of pesticides and chemical fertilizers in farming, and the lack of flavor and texture in most store-bought produce, and a trend is on the move. More and more people are concerned with food as the raw material to maintain their health and that of their families, and are looking towards growing their own as the most economic and reliable means to providing nutrient-rich, toxin-free food.

Naturally it is not only country people but also town and city dwellers who want to make the most of their plots of land to grow food for their tables and herbs both for their kitchens and their medicine chests. Many people do not possess sufficient space to grow their fruits, vegetables, and herbs in neat rows or stylish potagers set apart from their lawns and flower borders, so it has become increasingly necessary to plan the plot of land to maximize productivity. Indeed, garden fashions have changed, and today gardens are expected to be places of both beauty and productivity, the aesthetic value of the garden being equal to that of the food that can be harvested from it. It requires a thoughtful design to provide eye-catching appeal as well as a steady flow of food from the garden for as much of the year as possible.

This book contains a variety of traditional planting designs and layouts which may be useful as springboards for your own creations, and which could be adapted to your particular plot, its size and exposure.

Medieval monastic vegetable gardens were laid out to make the maximum use of limited space in small square or rectangular beds. Vegetables, fruits, and herbs were usually planted in neat rows, with room to walk between them to weed and maintain them. In the medieval physic gardens, beds for medicinal herbs would have been of similar design, with an enclosed rectangle or square divided into small rectangular raised beds edged with elm planks. Raised beds aid drainage on heavy, wet soils and enable organic compost to be added at regular intervals to improve soil fertility. This system can be followed today and looks very attractive. It is easy to maintain, and allows you to create a variety of soil conditions in different areas of the garden, making it possible to grow a wide range of produce.

Narrow paths divided the raised beds. These enabled planting and maintenance of the beds and at the same time defined a geometric design which was pleasing to the eye and had symbolic value. Often the monks created a cross in the center of the garden—the symbol of Christianity—which remains popular to this day.

In vegetable and herb gardens, different produce was grown in each space defined by the paths. This allowed a simple rotation plan, which could easily be adapted for gardens today. The paths were interspersed with pergolas and arbors to support climbers such as roses, hops, and vines and as places for tranquillity and contemplation.

Until the 19th century, an English cottager's garden was designed for food production rather than aesthetic value. It was usually small, and cottagers would crowd in as many flowers and herbs as they could. The cottage garden was traditionally laid out with a straight path leading to the front door, and a boundary hedge or wall with a gate—possibly with an arch over the gate or trellis by the front door supporting climbing roses or honeysuckle. To either side would be neat rows of vegetables and soft fruit edged with flowers and herbs, or there would be a larger plot around the back for such purposes. Almost anything goes in a cottage garden: an area outside the kitchen for herbs, for example, or walls to support espaliered fruit trees. The advantage of a cottage garden design is that it does not require as much maintenance as a more formal garden, and involves a more relaxed approach to gardening. A broad range of vegetables, flowers, herbs, and fruits can be grown together creating a medley of color, shape, and produce which is enhanced by a sense of profusion rather than order.

The potager garden is a more decorative, stylized version of a kitchen garden, designed to delight the eye as well as the palate. The potager originated in France where vegetables, fruits, herbs, and flowers are often arranged in attractive, symmetrical patterns in a formal layout. Today this is an increasingly popular breakaway from the traditional kitchen garden with its straight rows and allows for greater creativity and individual design using color, texture, shape, and form to provide a highly decorative tapestry of food.

The success of potagers depends on their design, and consideration of the size, shape, and color of the plants used. The width and height of each species needs to be taken into consideration, just as much as the color and texture. Low-growing species should be placed in front of taller ones so that they, too, can have a share of the sun. Flowering annuals— such as calendulas and nasturtiums— can provide vibrant color not only in the garden but also in the salad bowl. Tall plants for the back could include rhubarb, swiss chard, peas, tomatoes, broccoli, brussels sprouts, and pole beans trained onto a trellis or tripod. For the middle or front, dwarf beans, beet, carrots, celery, chives, basil, cabbage, parsley, onions, radishes, and turnips. As the vegetables are harvested, the beautiful pattern will start to disappear, although they can always be harvested in a symmetrical fashion.

Alternatively, for those with a minimal amount of space and/or time available, herbs, vegetables, and fruits can all be grown together in a flower border. This looks attractive and it also provides instant companion planting. The border needs to be well located in a sunny position with good rich soil to provide nutrients for such a range of plants to grow so intensively. Vegetables and herbs can be grown in the spaces in between shrubs and flowers; fruit trees will provide beautiful blossoms in springtime and fruit in late summer or early autumn, and fruit bushes can be grown as deciduous hedges to provide shelter for the garden. Here again, taller varieties of vegetables should be grown at the back of the bed, along with "large architectural" herbs such as fennel, borage, angelica, or horse-radish. Low-growing thyme and chamomile could be placed at the front.

Some like to grow their culinary herbs separately in a herb garden, often just near the kitchen. This needs to be a warm sheltered place with well-drained soil to produce the best results. With all the variety of shapes, textures, colors, not to mention smells, the herb garden can be a place to delight not only the eye and the palate but also the nose—so it is best to place the herb garden where it can be enjoyed for all its virtues. Pots of herbs can be positioned on patios, verandas, or next to the front door or window. Many herbs grow quickly into clumps or shrubs and need to planted with plenty of space between them to allow for growth. Any spaces can be filled with annual herbs such as nasturtiums, calendulas, basil, and coriander. Small paths through the herb garden will allow for maintenance and easy harvesting.

Even since Roman times, city dwellers have grown herbs and other plants in terracotta pots and containers on balconies and windowsills. For those who have very limited space, herbs can be grown in pots and containers along with a few salad vegetables, such as chicory, arugula, looseleaf lettuce or Romaine lettuce. Basil, coriander, parsley, chives, and oregano can be grown alongside these. Pots look highly attractive and allow some of the more rampant herbs—such as tarragon, marjoram, chives, mint, or lemon balm—to be nicely contained.

Everyone can have some kind of kitchen garden, even if it consists only of a few pots of herbs or salad vegetables on the kitchen windowsill, or on the balcony, peas and beans trailing between the railings, and a hanging basket of tomatoes on the wall—all can provide fresh, natural food that can be harvested and eaten within minutes, bringing an immense feeling of achievement and satisfaction. Nurturing a seed to its full potential as a food or medicine is a truly rewarding experience.

the healing garden

'The lesson I have thoroughly learnt and wish

to pass on to others, is to know the enduring

happiness that the love of a

garden gives.'

Gertrude Jekyll

The pleasures of gardening

There is something about gardens and gardening that is almost compelling. When growing edible and health-giving plants as well as flowers, it is possible to create a place not only of beauty but also of purpose. A garden can be a place in which to sit and relax, and it can also be a place to put your heart and your back into and reap the rewards.

There is immense satisfaction to be gained by tending a garden, and perhaps even more from growing your own food. Not many pursuits in life are able to keep you fit, remove stress, lift your spirits, and produce a wealth of nutrient-packed and toxin-free fruits, herbs, and vegetables. Organic kitchen gardening can do just that!

It is wonderfully rewarding to spend a few relaxed hours weeding, creating a sense of order and harmony in a previously untidy, overgrown area of the garden. Certainly, some of the challenges when maintaining a multipurpose garden include keeping order amid the range of different kinds of plants, keeping the balance between beauty and productivity, and finding combinations of plants and planting designs that work well for you.

Working in a an organic garden provides an opportunity not only to derive food, sustenance, and satisfaction for ourselves, but also to do something positive for the planet on which we live and on which we depend for our existence. We need trees to give us oxygen; grass, and scrub to prevent soil erosion; woodlands to preserve topsoil against winds and rains; and the wild herbs, fruits, and vegetables that originally provided us with the cultivated varieties on which we depend for our foods today.

Our ancestors have grown cultivated forms of vegetables for thousands of years and the food we grow in our gardens today allows us to benefit from years of evolution, selection, and experiment, yet recent "progress" has proved far from beneficial. Pesticides and fertilizers may enable us to grow plenty of fruit and vegetables that look perfect and inviting but at what cost to the earth's ecosystem—and to our health? The

mass-market production of many food crops has taken a worrisome direction: genetic engineering can produce food with prolonged shelf life, bright colors, and a healthy appearance, but it is at the cost of its flavor and nutritional value. If a food has been transported for a long distance in cold storage, it may look fresh on the supermarket shelf, but its vitamin C and A content will not compare with that of a fruit or vegetable grown at home and eaten fresh from the garden.

Many gardeners are fast coming to realize both the drawbacks of chemical fertilizers and that insecticides not only destroy pests but also the beneficial organisms that keep other pests at bay. They know that in upsetting the delicate natural balance of their garden, they may well be paving the way for more serious problems ahead. As a result, they are looking for more ecological ways of gardening that use old-fashioned methods of husbandry in conjunction with modern methods of organic gardening—an agreeable blend of the old and the new. The vast inherited pool of garden folklore, from which we have learned methods of gardening such as pruning and propagating, still works well, while modern, disease-resistant plant varieties can eliminate the need for insecticides and chemical fertilizers, and new dwarf forms of vegetables help those of us who have little growing space.

Gardening provides constant stimulation and challenge, for the garden is always changing, and each garden has a life of its own. For our part, we need to work in conjunction with the natural forces in our garden, to our mutual benefit, rather than fighting to control nature with pesticides, insecticides, and chemical fertilizers. The give and take in the garden involves working with the end result in mind while taking into consideration what the garden itself needs to enable it to produce to maximum efficiency. Enhancing nature requires treating the garden with care and consideration, and acquiring some knowledge about the needs of the soil and the requirements of the plants you want to grow. Information of this kind is widely available in literature, but only experience will help you to store it in your memory. Experience comes from contact with the garden and with the soil through our five senses, and this physical relationship with the natural world will have a beneficial healing effect long before we enjoy the fruits of our labors at the dining table, and learn to appreciate their contribution to health and well-being.

The joy of seeing a beautifully laid out and well-maintained garden, combined with the colors and shapes of the vegetables, fruits, flowers, and herbs, is hard to describe. The sounds in the garden can quiet the mind, and connect us to the world of nature around us: bees humming, birds singing, and the rustle of the leaves in the breeze. The varying textures of plants provide much interest: marigolds are full of resins and gums which make the whole plant sticky to the touch, borage is coarse and hairy, while dill and fennel are light and feathery. Perfumes of plants have the ability to alter mood, awaken the senses, and release tension, all within a few minutes. Lemon balm, for example, has an uplifting and calming effect when its scent is inhaled, and has been used for centuries by herbalists for this purpose.

Finally, taste—there is no question that healthy fruits, vegetables, and herbs grown organically and picked minutes before eating have a freshness and flavor that can never be equaled by those you buy. This is one of the main reasons why sitting down to a meal containing ingredients that you have grown yourself is so satisfying. Food that is freshly picked from your own garden is packed with vitamins, minerals, and trace elements which have had no chance to diminish before they are eaten, and you can be

completely certain that they contain no chemical fertilizers or pesticides.

In addition to all the other considerations, there are several ways in which edible plants can benefit our health directly. First, they are a source of a wide range of nutrients that provide our bodies with the building blocks for making new cells, repairing damage, and fighting off disease. Second, their cellulose provides fiber which, because it is not broken down in the bowel, helps to maintain a healthy bowel. Finally, they contain a range of pharmacologically active constituents such as mucilage, essential oils, antioxidants, and phytosterols which have specifically therapeutic effects. Peppers, carrots, parsley, and dandelion leaves, for example, are rich in the antioxidants beta-carotene and vitamin C, which help to delay the aging process, enhance immunity, and are thought to help in the prevention of heart and arter-

ial disease and some cancers. We are becoming increasingly aware from the wealth of modern research and media coverage that diet plays an enormous part in maintaining health and preventing diseases. But this knowledge is nothing new; for thousands of years cultures all over the world have utilized food to prevent as well as to treat illness.

It is reassuring to observe that while fashions in diet change from one decade to another, modern research is bearing out many of the ancient medicinal uses of fruits, vegetables, and herbs. Today we know that leeks, valued as remedies for the throat and chest by the ancient Greeks and Romans, contain, like their relatives garlic and onions, antiseptic substances that help ward off respiratory infections, colds, flu, sore throats, and chest infections, while the mucilage they contain helps soothe inflammatory conditions such as laryngitis. Turnips were

also valued highly by the Greeks and Romans and grown in medieval monasteries by monks who considered them beneficial to the stomach and to "dry" intestines. Today we know that the fiber in turnips has beneficial effects on the bowel, helping to ensure normal function, preventing constipation, and lowering the risk of bowel disease, including cancer. There are other foods that have not yet been researched but for which several thousand years of human use clearly indicate therapeutic uses. Moreover, the fact that humans have been eating plants and using herbs as medicines for so long means that our bodies have adapted to natural foods in a way that they have not always done to the synthetic components found in many modern medicines.

With such a wonderful array of therapeutic properties—the healing benefits of actually tending a garden and creating a place of beauty, and its productivity— the kitchen garden can be proudly displayed, not hidden from view as it might have been fifty or a hundred years ago. Dreary rows of cabbages and sprouts, interspersed with drab-looking potatoes, are a thing of the past. Today, a profusion of healthy-looking vegetables in all shades from red to green to yellow, intermingled with the bright flowers of medicinal herbs such as nasturtiums, borage, and marigolds, and blossoms of fruit trees in spring followed by an abundance of fruit glowing in autumnal sunlight, conjures up a vision of the Garden of Eden—a place of solace, comfort, joy, a wonderful antidote to our stressful lives, a "good health garden" indeed.

what is paradise?
but a garden, an
orchard of trees and
herbs full of pleasure
and nothing there
but delights.
William Lawson

NUTRIENTS CONTAINED IN FOODS

Carbohydrates provide our main source of energy. Simple carbohydrates come in the form of sugars such as glucose and fructose, which occur in fruit. Complex carbohydrates are made up of several sugars and are found in starchy foods such as potatoes, beans, parsnips, turnips, carrots and pumpkins, as well as pasta and bread. When digested, complex carbohydrates are broken down into simple sugars, and require insulin from the pancreas to utilize them in the cells for energy. Such starchy foods also provide other nutrients—minerals, vitamins, and trace elements—and diets in which starchy foods supply around half the daily calorie intake have been shown to help prevent heart and arterial disease, strokes, obesity, gallstones, and bowel problems. Calories eaten as starch are less likely to be stored as body fat than calories eaten in the form of fats, and so help to prevent overweight.

Soluble fiber is found in many fruits and vegetables and has the effect of slowing the rate of digestion in the stomach and intestine, so providing a steady flow of energy as the food is digested and absorbed. Sugars and low-fiber starchy foods are digested and absorbed more rapidly than those containing soluble fiber, which means that blood sugar levels rise quickly, with the help of insulin from the pancreas, but also fall fast. Soluble fiber helps to produce a lower rate of rise and fall in blood sugar, something that is vital to diabetics and those who suffer from hypoglycemia. It also helps to stabilize energy levels and mood. Diets rich in soluble fiber such as oat bran

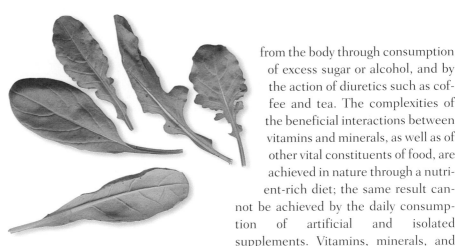

have been shown to reduce low-density lipoprotein (LDL) cholesterol, the "bad cholesterol" which contributes to heart and arterial disease.

Fats in plants are found in the form of fatty acids, which are essential for the absorption of vitamins A, D, E, and K from the diet. Fatty acids are vital to normal function of the nervous, immune, and hormonal systems, and help maintain the health of the heart and circulatory system, thereby helping to prevent heart disease and strokes.

Proteins are necessary for the formation and repair of cells and as building blocks for hormones and enzymes. Proteins are synthesized in the body from amino acids derived from food. Complete proteins, containing all the essential amino acids, are found in animal proteins—fish, poultry, meat, eggs, and milk. No vegetable proteins except soybean products such as tofu and tempeh contain all the essential amino acids, and therefore two vegetable proteins should be eaten together at the same meal. Beans and other legumes are good sources of protein but must be combined with one of the other two main sources of vegetable protein: nuts and seeds or grains.

Vitamins and minerals are also vital to health. Fifteen vitamins and fifteen minerals can only be obtained from food to help prevent disease. Many are leached

from the body through consumption of excess sugar or alcohol, and by the action of diuretics such as coffee and tea. The complexities of the beneficial interactions between vitamins and minerals, as well as of other vital constituents of food, are achieved in nature through a nutrient-rich diet; the same result cannot be achieved by the daily consumption of artificial and isolated supplements. Vitamins, minerals, and trace elements can also be lost through cooking and storing fruits and vegetables. Produce is best eaten straight from the garden, either raw, lightly steamed, or stir-fried. Boiling in large quantities of water and throwing the cooking water away will result in considerable loss of nutrients.

Antioxidants are present in many foods, and these have the ability to prevent oxidation in the body, which causes the release of harmful substances known as free radicals. Free radicals contribute to heart disease, cancer, degenerative diseases, lowered immunity, cataracts, and the aging process. Vitamins A, C, and E, selenium, and many carotenes and flavonoids from food sources have been shown to exert a beneficial effect in a way that chemical supplements of these substances do not. Zinc, copper, and selenium deficiencies reduce antioxidant defenses, and excess iron increases oxidation. Red and orange fruits and vegetables (red and yellow peppers, tomatoes, carrots, oranges, strawberries, raspberries) and spinach are good sources of the antioxidants beta-carotene, flavonoids, and vitamins C and E.

Carotenes give the color to orange, red, and yellow foods; they number in the hundreds, and act as antioxidants. Studies have shown that a carotene-rich diet lowers incidences of heart and arterial disease, cataracts, and some cancers.

Popular vegetables such as carrots, onions, potatoes, and garlic are rich in essential vitamins, minerals, and trace elements,

vegetables

and helpful in bolstering our resistance to disease. In addition, studies have shown that these vegetables can be medically beneficial in the treatment of a number of ailments including arthritis, heart disease, and digestive problems.

Onion *Allium cepa*

The onion has been used in healing since 4000 B.C. Its benefits were first appreciated by the ancient Egyptians, who respected it both as a panacea and as a symbol of vitality. For centuries, it has been used to ward off epidemics of infectious diseases such as typhoid, cholera, and the plague, as well as to increase energy and longevity, and modern medical research has served to justify our ancestors' belief in this amazing plant.

ONIONS CAN HELP TREAT

- *Anemia*
- *Arthritis*
- *Bronchitis*
- *Congestion, coughs, and colds*
- *Constipation*
- *Cystitis*
- *Flatulence*
- *Flu*
- *Fluid retention*
- *Gout*
- *High blood pressure*
- *High cholesterol*
- *Pharyngitis*
- *Rhinitis*
- *Sinusitis*
- *Sore throat*
- *Worms*

INTERNAL USE

Raw onions are antiseptic, and they enhance the body's ability to fight against a long list of bacteria, including salmonella. They are also effective against respiratory infections and can be used in the treatment of gastrointestinal infections and urinary tract infections such as cystitis. The pungent onion has a heating effect, increasing the circulation and causing sweating, which is useful in cold, damp weather to ward off infection, and helps to lower fevers and sweat out colds and flu. Onions can also be used to treat sore throats, congestion, and sinusitis since they have a decongestant action. They act as an expectorant for coughs and bronchitis.

Raw onions act as a stimulant to the digestive system and liver, and thereby make an excellent nutritional energy tonic since they enhance the digestion and absorption of nutrients. Thus, they can also be used as a pick-me-up, and when recovering from illness. Onions have earned a reputation as a blood purifier and detoxifying remedy. They also have diuretic properties and can be used for water retention, arthritis, and gout. A few tablespoons of onion wine taken first thing each morning for 8 to 10 days is an excellent remedy for worms.

Traditionally, both onions and garlic have been used to benefit the heart and circulation, and they can help to reduce blood pressure and narrowing of arteries, and to guard against heart attacks. Modern research has shown that they have a lowering effect on harmful cholesterol, and half a raw onion eaten daily is said to significantly lessen the chance of a heart attack. Both raw and cooked onions help to lower blood pressure, and prevent clots. Onions have been used in folk medicine to reduce blood sugar levels, which could be useful to diabetics, and research has shown that both raw and cooked onions do, in fact, have this effect. It is also thought that, due to their sulfur compounds, they may be useful as a remedy against cancer.

EXTERNAL USE

Onions have many external applications: for example, for wasp and bee stings, rub a slice of raw onion over the affected area. For warts, chop onions, cover them with salt, and leave them overnight. Store the juice that collects in a bottle, and dab the warts with it twice daily. Onion juice can also be applied to burns or as an antiseptic to cuts and abrasions, and used for toothache—simply place a cotton swab soaked in onion juice into the cavity. Poultices of cooked onion paste applied to abscesses and boils can give speedy relief. Grated onions can be made into a poultice to apply to chilblains.

When the leaf tips turn yellow, bend the leaves over at right angles and slightly loosen the roots with a fork.

HOW TO GROW

Seeds are obtainable from garden centers but onions are easiest raised from "sets." Add well-rotted manure before planting, and extra lime if your soil is acid. Sow seeds in late winter or early spring under glass or in a warm bed in spring. Harden off seedlings started indoors and plant out seedlings or sets in midspring, 4 in (10 cm) apart, with 10 in (25 cm) between rows. Mulch in the summer. Dig up onions in early autumn after loosening the roots and spread them in the sun to dry the foliage.

Onion
recipes & remedies

Any type of onion, including scallions, can be used to make the following remedies.

Onion wine

A useful remedy to expel worms from the intestines.

1 large onion, finely chopped
4 tablespoons (60 ml) honey
2 cups (570 ml) white wine

Add the onion and honey to the wine, and let it stand for 48 hours, shaking frequently. Strain, and take 2–4 tablespoons (30–60 ml) daily. Will keep for up to 3 days if it is refrigerated.

Infusion of onions

For colds, congestion, coughs, and bronchitis.

4 medium-sized onions
3¹/2 cups (1 liter) hot water

Slice the onions into the hot water and soak for 2–3 hours. Take one glassful twice daily.

Onion decoction

A refreshing drink that is particularly helpful for congestion.

3 onions, cut into quarters
3/4 cup (250 ml) water
Honey to taste

Simmer the onions in water for 5–10 minutes. Strain, add honey to taste and drink as a beverage.

Onion skin tea

Tea made from boiled onion skins can help to improve poor circulation, especially in cases of gout.

10 onion skins
3¹/2 cups (1 liter) water

Simmer the onion skins in the water for 10 minutes. This quantity may be drunk each day for as long as symptoms persist.

Onion juice

For colds, congestion, coughs, and bronchitis.

2 medium-sized onions
2 tablespoons honey

Chop the raw onion into a bowl, drizzle with honey and let stand, covered, at room temperature overnight to produce a juice. Take 1 tablespoon (15 ml) every 2 hours.

Leek *Allium porrum*

The leek has been praised for its medicinal benefits for thousands of years. The ancient Greeks recommended leeks to improve the voice, and for a few days every month, the Roman Emperor Nero apparently ate only leeks and oil, to keep his throat healthy in order to sing well during his competitive appearances at festivals. The leek has been a popular remedy for the respiratory system ever since.

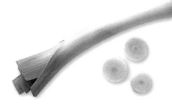

LEEKS CAN
HELP TREAT

- *Atherosclerosis*
- *Boils and abscesses*
- *Bowel infections*
- *Chest infections*
- *Colic*
- *Cuts and abrasions*
- *Congestion, coughs, and colds*
- *Diarrhea*
- *Flatulence*
- *Insect bites and stings*
- *Minor burns and scalds*
- *Poor circulation*
- *Sore throats*
- *Urinary infections*

INTERNAL USE
Leeks remain an effective medicine for the respiratory tract. They have warming and stimulating properties that help to loosen phlegm, and an expectorant action that helps to clear congestion from the nose and bronchial system. In addition, leeks, like other members of the *Allium* family, have antiseptic properties which help to ward off colds, sore throats, and chest infections. These warming and antiseptic properties can also be put to good use in the digestive system. Leek broth, leek soup, or braised leeks can be taken to ease diarrhea. Leeks can help to reestablish the normal bacterial population in the gut after a bowel infection, during and after a course of antibiotics, or when suffering from candidiasis. They also have a relaxing effect in the digestive tract, helping to relieve cramp, colic, gas, and distension. They stimulate the appetite and improve digestion, and by aiding peristalsis, are mildly laxative. The mucilage in leeks has a soothing action, which can help to relieve inflammatory conditions such as gastritis and colitis.

Leeks have a diuretic action, enhancing the elimination of fluid and wastes via the urinary tract. The combined diuretic and antiseptic properties can be useful when treating fluid retention, or urinary infections such as urethritis and cystitis.

The leek is rich in potassium, magnesium, iron, and calcium, as well as in the vitamins C and B complex. Its warming properties stimulate the circulation and are generally invigorating, and since leeks are nutritious and easily digested, they make an excellent tonic for convalescence. As a broth, leeks provide a good nutrient-rich food to replace minerals lost during sickness and diarrhea. Like onions and garlic, leeks help to protect the heart and arteries and to prevent atherosclerosis and high blood pressure.

EXTERNAL USE
A leek poultice can be used as an antiseptic for cuts and abrasions. A cut leek can be rubbed on to insect bites and stings to relieve pain and swelling. Warm cooked leeks can be mashed to make a paste, which, when applied on gauze, can help to draw poisons out of boils and abscesses. Leeks cooked in milk make a soothing and healing lotion to apply to inflamed skin and minor burns.

HOW TO GROW
Leeks like rich, well-drained soil with plenty of organic matter. The seeds should be germinated indoors in late winter or outdoors in spring. If they are outdoors in a seed bed, sow the seeds about 1/2 in (1.5 cm) apart, in rows 6 in (15 cm) apart in late spring. Plant them out in their cropping positions between mid and later summer when 8 in (20 cm) high. Make holes in firm ground 8 in (20 cm) deep and 6-8 in (15–20 cm) apart in staggered rows. Drop the leeks in gently, roots downward, then fill with water to settle the soil around the roots. The seedlings will soon settle without the soil being replaced. Leeks can be planted out in the space left by early potatoes. They should be "earthed up" periodically *(see left)*, hoed in summer to keep the weeds down, and kept watered during dry weather. They can be harvested from early winter to midspring.

To "earth up," draw in soil around the base of the leek with a hoe in order to prevent light getting to the plant. This will increase the length of the white stem.

Leek
recipes & remedies

Young leeks can be eaten raw as part of a salad. However, they are usually served steamed or braised.

Leek broth

A good mineral-rich food to take when suffering from cystitis, diarrhea, or gastroenteritis.

8–10 leeks, chopped
7–10 cups (2–3 liters) water

Simmer the leeks gently in water for 1¹/₂–2 hours. Serve the broth with the leeks intact or pass them through a blender first, depending on preference.

Syrup of leeks

Good for sore throats, congestion, colds, and chest infections.

2–3 leeks, chopped
3¹/₂ cups (1 liter) water
Honey to taste

Cook the leeks in the water until soft and extract the juice by squeezing through a cheesecloth. Add honey to taste and give 1 teaspoon (5 ml) 3–6 times daily for respiratory infections. Store in the refrigerator.

Leek poultice

A useful antiseptic for cuts and abrasions.

1 leek, washed and chopped
3¹/₂–7 cups (1–2 liters) water

Cook the leek gently in water until soft. When cool enough to handle, place it between 2 pieces of gauze and bind it to the affected part with a cotton bandage.

Leek, garlic, and ginger soup

A warming soup, helpful for warding off colds, flu, sore throats, and chest infections, and for clearing congestion.

4 medium-sized or 5 small leeks
1 tablespoon olive oil
2 cloves garlic, minced
1 tablespoon chopped fresh ginger
1 quart (1.2 liter) water or vegetable stock
Salt and freshly ground pepper

Trim the leeks and finely slice them. Heat the oil in a large saucepan over low heat. Add the leeks, cover, and cook for 10 minutes, or until just softened. Add the garlic, ginger, and water or stock, increase the heat, and bring to a boil. Reduce the heat, cover the pan, and simmer the soup for 20 minutes, or until the leeks are tender. Remove the pan from the heat and allow the soup to cool a little, then puree in a food processor or blender until smooth. Return to the rinsed-out saucepan and reheat gently. Season to taste with salt and pepper, and serve.

Garlic *Allium sativum*

Garlic is one of the most remarkable food or herb remedies in existence, and has been revered as a rejuvenator and aphrodisiac in many cultures for thousands of years, despite its antisocial effects. The ancient Greeks sang its praises, and the Egyptian pharaoh Cheops was well aware of its energy-giving properties when he ordered the workmen building the Great Pyramid to have a daily ration of garlic; this was designed not only to give them strength but also to protect them against epidemics, as it has long been known for its ability to fight infection.

GARLIC CAN HELP TREAT

- *Asthma*
- *Chest infections*
- *Chilblains*
- *Circulatory disease*
- *Congestion, coughs, and colds*
- *High blood pressure*
- *High cholesterol*
- *Poor circulation*
- *Sore throats*
- *Tiredness and lethargy*
- *Worms*

INTERNAL USE

As a result of garlic's invigorating properties, it has been considered the vital ingredient of many an elixir of life. In 5th-century Greece it was apparently sold by street vendors chanting: "It is truth. Garlic gives man youth." They were not far wrong, because modern research has verified garlic's rejuvenating properties by revealing the presence of antioxidant substances that help to delay the aging process by protecting the body against damage from free radicals. It has also been found that garlic's antioxidant substances help to guard against degenerative diseases such as heart disease and cancer. The sulfur compounds contained in garlic are reported to slow down the growth of tumors.

Garlic is known to have powerful antibacterial, antifungal, antiviral, and antiparasitic properties, and when it is crushed it not only releases its powerful odor but also its natural antibacterial substance, allicin, which has been shown to have antibiotic properties. When taken internally, garlic exerts its antibacterial effects throughout the digestive, respiratory, urinary, and reproductive tracts. It makes an excellent remedy for sore throats and colds. It helps to re-establish the normal bacterial population of the gut after taking orthodox antibiotics for an infection. It is an effective remedy for worms when taken first thing in the morning on an empty stomach. Its warming and stimulating properties have a decongestant action in the respiratory system, helping to clear congestion and sinusitis. It also acts as an expectorant and can be helpful to clear coughs and chest infections. Because of its diaphoretic properties, garlic helps to bring down fevers. Garlic has a warming and invigorating effect on the digestion, stimulating the secretion of digestive enzymes and of bile from the liver. It promotes appetite and improves the digestion and absorption of food, helping to ensure regular bowel movements and thereby keeping the bowel free of excess toxins.

Garlic has been valued in India as a remedy for the heart and circulation since the first century A.D., and in China and Japan it has been taken for centuries to lower blood pressure. Modern research has again confirmed its ancient use, showing that, if taken regularly, garlic can significantly lower the level of harmful cholesterol, thereby helping to prevent and treat atherosclerosis and high blood pressure. By reducing blood pressure and the tendency to clotting, garlic helps to prevent heart attacks and strokes. It has a vasodilatory action, opening the blood vessels and improving blood flow through them, helping to relieve and prevent a wide range of circulatory disorders and promoting a feeling of warmth and wellbeing.

When harvesting, ease the plants out of the ground with a fork to avoid damaging them.

HOW TO GROW

Garlic does best in light, well-drained, well-manured soil in a sunny position. It can be propagated by planting in late winter to early spring, pointed end upwards, about 6 in (15 cm) apart. Harvest in early to mid autumn, after the foliage has died down. Allow the plants to dry in the sun before tying them into bunches and storing them in a cool, dry place.

Garlic
recipes & remedies

To be beneficial, garlic needs to be eaten raw, since cooking destroys about 95% of its medicinal value.

Garlic juice

Excellent for coughs, colds, and sore throats

4 garlic cloves, sliced thinly
Honey to cover

Cover the sliced garlic with honey and let stand for 2–3 hours. Crush to extract the juice, and take teaspoonfuls throughout the day.

Garlic oil

For rheumatism, arthritis, sprains, strains, and chest infections.

6 garlic cloves, peeled and crushed
3/4 cup (250 ml) olive oil

Place the cloves in a sterilized bottle and cover with olive oil. Seal and let stand for 1–2 weeks. Press through cheesecloth and store the oil in a sterilized bottle. Massage the oil regularly into affected areas of the body.

Celery *Apium graveolens* var. *dulce*

Celery is a tasty vegetable with a distinct, pungent flavor, and makes a crisp addition to salads. The celery we grow today is the descendant of wild celery (*A. graveolens*), which was highly valued by the ancient Egyptians, Greeks, and Chinese, both as a flavoring and as a medicine. The Romans used to wear a wreath of celery around their heads to ease a hangover. In medieval times, celery was popular for its ability to relieve aches and pains, to calm the nerves and benefit the digestion.

CELERY CAN
HELP TREAT

- *Arthritis*
- *Constipation*
- *Gout*
- *High blood pressure*
- *High cholesterol*
- *Mild depression*
- *Stress*
- *Urinary infections*
- *Weak digestion*

INTERNAL USE

Celery has been valued as a remedy to lower blood pressure since about 200 B.C., and modern research has shown that eating a couple of stalks of fresh celery daily can help to reduce blood pressure and harmful cholesterol levels. The substance responsible for lowering blood pressure has been identified as 3-n-butylphthalide, which acts by blocking the enzyme that makes catecholamines, which are stress hormones. This substance dilates the arteries and reduces contraction in arterial muscles caused by stress, and is probably most useful for treating stress-related hypertension.

Both celery leaves and seeds have a mild diuretic effect on the body because they contain appreciable amounts of apiol in their volatile oil, which has a cleansing effect on the urinary system. Celery helps to enhance the elimination of excess fluid and toxins via the urinary system and it has long been used to treat gout and arthritis as well as urinary infections. While orthodox diuretics can leach potassium from the system unless they are combined with a potassium supplement, celery comes ready packed with a good dose of this mineral.

Modern research has shown that celery contains at least eight different compounds that may be effective against cancer, including substances that may have the ability to neutralize the effects of some carcinogens. Eating raw celery on a regular basis may also help to protect against stomach cancer.

Celery has always been popular with dieters because it is low in calories and high in fiber, while its diuretic action reduces excess fluid. The stalks are moderately nutritious, but the top leaves are more so since they contain more calcium, iron, potassium, and vitamins A and C than the stalks. The leaves should be included when making soups, casseroles, and salads. Celery seeds and stalks both have a stimulating effect on the digestive tract: celery seed tea makes an excellent after-dinner *digestif*. Celery seeds have also been used by herbalists for centuries to raise the spirits and benefit the nervous system. Celery, being high in fiber, will help to ensure regular bowel function and so prevent and relieve constipation.

Caution: Celery eaten either before or after vigorous exercise has been known to induce allergic responses in sensitive people.

HOW TO GROW

Celery needs a rich, moist, loamy soil if it is to do well and thrive. To prepare the celery beds, dig trenches in late winter, 1 ft (30 cm) deep, and fill them almost to the top with well-rotted manure or compost. Cover with a few inches of rich soil. Sow seeds in late winter, indoors or in a greenhouse, on fine soil. Cover seeds with ¼ in (6 mm) of fine potting mix. Keep the seeds well watered and be patient, since celery takes up to 3 weeks to germinate. Once the celery seedlings have reached about 1 in (2.5 cm) in height and have developed their first true leaves, transplant them into 4 in (10 cm) pots, but only one seedling per pot.

Move the young plants into their prepared bed in late spring, when there is no danger from frosts. Water them well, and keep them watered especially in dry weather. Side-dress the plants once a month using well-rotted compost. As the celery plants develop, mound up the soil around them to keep the stems white and tender. Individual stems can be picked as required, or the whole plant can be harvested by cutting it off at the roots.

Celery
recipes & remedies

Tender young hearts are delicious raw, and stalks are often eaten with a cheese filling. They make a good alternative to bread or crackers as a base for hors-d'oeuvres or snacks, when they can be stuffed with cream cheese.

Celery seed tea

An excellent after-dinner drink to aid digestion.

2 oz (50 g) celery seeds
2 cups (570 ml) boiling water

Add the seeds to the water. Cover the pan and let infuse for 5–10 minutes. Strain and allow the liquid to cool. Drink 1 cupful after meals.

Celery juice

Drink half a glass first thing in the morning for 15–20 days as a remedy for rheumatic ailments.

4–6 celery stalks

Remove the leaves and wash the stalks thoroughly, then extract the juice using a juicer.

Horseradish *Armoracia rusticana*

Horseradish is probably native to southeast Europe but is widely cultivated all over northern Europe and North America, and is often found growing wild on banks and roadsides. It is popular for its large taproots, which make an excellent, intensely pungent condiment. Freshly grated horseradish is traditionally eaten as horseradish sauce with beef and fish, often mixed with grated apple and cream or yogurt to reduce its biting pungency.

HORSERADISH CAN HELP TREAT

- *Arthritis*
- *Boils and abscesses*
- *Bronchial congestion*
- *Chilblains*
- *Congestion, coughs, and colds*
- *Constipation*
- *Flu and fevers*
- *Fluid retention*
- *Gout*
- *Hay fever*
- *Indigestion*
- *Poor appetite*
- *Poor circulation*
- *Sinusitis*
- *Skin problems*
- *Tiredness and lethargy*

The leaves of horseradish look like large dock leaves and also have a very pungent taste—when young and tender they can be chopped and mixed (in moderation) in green salads. Both the leaf and root are rich in vitamin C.

INTERNAL USE

Horseradish has been respected as a food and medicine since at least the time of the Romans, and was one of the five bitter herbs eaten by the Jews at the feast of the Passover. In medieval Europe it was taken to improve digestion and, in the 17th century, Culpeper prescribed it for external use to treat sciatica, gout, joint pain, and "hard swellings of the spleen and liver." It was not until the mid-17th century that it became popular as a condiment in Britain, served with meat to ease its digestion.

Horseradish is a powerful and stimulating remedy, increasing circulation and promoting warmth throughout the body—excellent for those who feel the cold and suffer from poor circulation in winter. Its pungency has a beneficial effect throughout the digestive tract when used in small amounts. It enhances appetite and stimulates digestive juices and bile, thereby aiding digestion. It helps to ensure regular bowel function and the movement of wastes through the system. By increasing blood flow to the tissues and removal of waste products from the body, horseradish acts as a good detoxifying substance. These qualities are augmented by its diuretic action (provided mainly by its constituent asparagin), which hastens elimination of fluid and toxins from the body. So, horseradish has often been used to cleanse the body, to clear the skin, and to treat problems such as boils and abscesses as well as arthritis and gout.

In the respiratory system, horseradish has a stimulating decongestant and expectorant action, and horseradish syrup has long been used to help relieve coughs, colds, fevers, flu, congestion, sinusitis, and hay fever. The powerful antibiotic effect it has is also excellent for urinary and respiratory infections. In addition, horseradish makes a good energy tonic.

EXTERNAL USE

A poultice of grated raw horseradish stimulates the circulation and has long been a folk remedy for easing arthritic pain. When applied, it should not come into direct contact with the skin as it can cause irritation and even blistering.

Caution: Horseradish should not be used for symptoms characterized by heat—such as acidity, gastritis, and peptic ulcers—nor by people with kidney or thyroid problems.

Horseradish should be deadheaded regularly to ensure continued flowering, and the roots dug up from autumn onward.

HOW TO GROW

Horseradish likes well-drained soil and prefers full sun, although it will tolerate partial shade. It grows up to 20 in (50 cm) tall. Propagate it either by sowing seeds in early spring, or by planting roots, which can be bought or dug up in the autumn of the previous year and stored. Planting holes should be 2 ft (60 cm) apart, and 14 in (25 cm) deep for roots. Horseradish grown from a root should be left in the ground for 2 years. It self-seeds readily, and once it begins to spread it can be hard to get rid of.

Horseradish
recipes & remedies

Horseradish gives off fumes that are more powerful than onions, which can cause the eyes to smart and run while preparing it. One way to avoid this is to grate or shred it in a food processor, rather than by hand.

Horseradish vinegar

To stimulate the appetite and aid digestion.

1 large horseradish root, washed, peeled, and grated
1 onion, chopped
2 cups (570 ml) cider vinegar

When grating the horseradish root, take care not to get any juice in your eyes. Fill a bottle or jar with the horseradish and onion. Heat the vinegar in a pan until it just starts to simmer and pour over. Let it cool and cover. Strain and store in a sterilized bottle. Shake well and let it stand for 4–6 weeks, shaking a few times a day. Take 1 teaspoonful (5 ml) in hot water 3–6 times daily when the need arises.

Horseradish syrup

A remedy for hayfever, bronchitis, and coughs.

1 large horseradish root, washed, peeled, and grated
Honey to cover

Place the grated horseradish root in a bowl and pour the honey over. Cover the bowl and leave for 24 hours. Press through a sieve or cheesecloth and store juice in a sterilized airtight bottle in the refrigerator. Take 1 teaspoon (5 ml) 3–6 times daily when symptoms arise.

Horseradish sauce

An excellent warming digestive and circulatory stimulant which aids in digestion of rich foods and heavy meats. Serve with cold meat or as a dip for crudites.

1/4 cup (50 g) grated fresh horseradish
1 tablespoon Dijon mustard
1 teaspoon cider vinegar
Pinch of paprika
1/4 cup (60 ml) natural yogurt
1 teaspoon fresh lemon juice

Mix the horseradish, mustard, vinegar, and paprika together in a small bowl. Add the yogurt and lemon juice, and stir until smooth. Refrigerate until chilled.

Asparagus *Asparagus officinalis*

Asparagus is best known as a culinary delicacy, enjoyed particularly in spring when the succulent new shoots, known as "spears," lightly steamed and drizzled with warmed butter or olive oil, provide a tasty hors-d'oeuvre. What is not so well known, however, is that several varieties of asparagus have been valued as medicines for more than 2,000 years. The ancient Egyptians cultivated it, as did the Greeks, who recommended it for rheumatism and urinary problems. In medieval monasteries it was valued for its detoxifying properties.

ASPARAGUS CAN HELP TREAT

- *Arthritis*
- *Cataracts*
- *Constipation*
- *Fluid retention*
- *Skin problems*
- *Tiredness and lethargy*
- *Urinary infections*
- *Vitamin and mineral deficiency*
- *Worms*

INTERNAL USE

Today, *A. officinalis* is used principally for its powerful diuretic effect, brought about mainly by its active constituent, known as asparagine, which helps to hasten the elimination of toxins and excess fluid from the body. It is helpful for relieving urinary infections and it also has a beneficial cleansing effect on the system.

Asparagus was popular in the past in the treatment of chronic arthritis, rheumatism, and gout, as well as skin problems. The rhizomes have been used in decoctions for their even stronger diuretic effect.

The cleansing properties of asparagus are augmented by the beneficial effect that this vegetable is thought to have on the functioning of the intestines and the liver, the great detoxifying organ of the body. It also stimulates bowel function and makes an effective remedy for mild constipation. Its asparagusic acid has the ability to expel worms from the body and has been used to treat schistosomiasis, which is an inflammatory condition of the liver.

Asparagus not only tastes delicious but is also a nutritious food. It contains folic acid, vitamins A, B complex, and C, manganese, iron, phosphorus, and protein. It makes a good food for dieters, too, being high in fiber and low in calories (unless covered in olive oil or butter, of course). The antioxidant vitamins A and C help to prevent damage to the cells caused by free radicals, and thereby help to slow the aging process, perhaps explaining why asparagus has long had the reputation of being a rejuvenative and restorative.

Another antioxidant substance found in asparagus—glutathione—has recently been shown to help to prevent the formation of cataracts and it is thought to have anticancer properties, as well as helping to boost the body's immunity to viruses.

HOW TO GROW

Asparagus plants are perennials and require a permanent bed in a sunny, sheltered site, where they can remain productive for up to 20 years. They grow best in a rich, sandy, well-drained soil with a nutritious top dressing. On heavy or wet soils it is best to grow them in a raised bed.

Asparagus may be grown from 1- to 3-year-old crowns or from seed. Dig in plenty of garden compost or well-rotted manure the year before planting and remove perennial weeds. Crowns should be planted in spring, in a trench about 1 ft (30 cm) wide and 8 in (20 cm) deep, with a ridge in the middle. Plant crowns in holes 4 in (10 cm) deep covered with 2 in (5 cm) of sifted soil. Rows should be 12–20 in (30–45 cm) apart.

When planting asparagus crowns, spread the roots over the ridge you have made in the center of the trench before covering with soil.

If planting from seed, soak them for at least a day before sowing in warm soil in spring. They should be planted 12–18 in (30–45 cm) apart, in rows 18 in (45 cm) apart, and thinned to 6 in (15 cm) apart when they reach 6 in (15 cm).

Plants can be harvested in their second or third year, in late spring or early summer. Cut the asparagus spears 1–2 in (2.5–5 cm) below the level of the soil when they are 5–7 in (13–18 cm) high. In autumn, the foliage should be cut back to about 1–2 in (2.5–5 cm) from the ground.

Asparagus
recipes & remedies

Asparagus spears can be steamed or boiled in water until they are tender, and eaten. The water left over from the cooking can be drunk 2–3 times daily as a cleansing diuretic. Alternatively, it can be prepared as a decoction.

Asparagus decoction

Excellent for urinary problems.

1 oz (25 g) asparagus, chopped
2 cups (570 ml) water

Place the chopped asparagus in the water. Bring it to the boil and then let it simmer for 20 minutes. Strain and let it cool. Drink 1 cupful 2–3 times a day. Store in the refrigerator.

Hand and foot bath

Recommended for stimulating the liver.

2 oz (55 g) asparagus rhizomes, washed
2 oz (55 g) asparagus, crushed
3 1/2 cups (1 liter) water

Place the asparagus and rhizomes in the water, bring it to the boil, and simmer for 15 minutes. Strain and let the liquid cool to body temperature. Soak the feet for 8 minutes in the morning and the hands for 8 minutes in the evening.

Brassicas *Brassica oleracea* varieties

Botanically speaking, the brassicas (members of the Cruciferae family) include cabbage and related green crops, such as kale, broccoli, brussels sprouts, and cauliflower, as well as horseradish, kohlrabi, radishes, rape, sea kale, and turnips. Cabbage, kale, broccoli, brussels sprouts, and cauliflower all have very similar properties and they can be used medicinally almost interchangeably. All of these vegetables contain both nitrogen and sulfur compounds, which can create gas in some susceptible people. It is the sulfur content that you can smell when these brassicas are cooked for too long.

BRASSICAS CAN HELP TREAT

- *Alcoholism*
- *Anemia*
- *Arthritis*
- *Boils and abscesses*
- *Chilblains*
- *Colitis*
- *Constipation*
- *Coughs and colds*
- *Cuts and abrasions*
- *Gastritis*
- *Gout*
- *Indigestion and heartburn*
- *Liver problems*
- *Low immunity*
- *Minor burns and scalds*
- *Poor lactation*
- *Sinusitis*
- *Skin problems*
- *Varicose veins*
- *Vitamin and mineral deficiency*

CABBAGE *B.oleracea* var. *capitata*

The humble cabbage is one of the most ancient and treasured remedies in history. In one form or another, it has been cultivated for about 4,000 years, and has earned the reputation of being a panacea for all ills, with such names as "poor man's medicine chest" and "doctor of the poor."

INTERNAL USE

Today, eaten raw or lightly steamed, cabbages are a valuable source of vitamins, minerals, and trace elements. They contain vitamins A, B, C, and E and minerals calcium, sulfur, silica, magnesium, iodine, iron, and phosphorus. Their antioxidant vitamins help to protect the body against against degenerative diseases including cancer, and to slow the effects of aging. Being rich in iron and chlorophyll, they make a good remedy for anemia. They have a reputation as a nerve tonic, and can be used to treat anxiety, insomnia, depression, and exhaustion. Cabbages were used by sailors to prevent scurvy and were recommended for pregnant women and breast-feeding mothers to increase milk production.

Cabbage has an ancient reputation as a remedy for purifying the blood, and still today, a decoction of raw cabbage taken daily makes a good cleanser to detoxify the system and it can be helpful for clearing skin problems such as acne and boils. It can also act as a diuretic, has-

tening the elimination of toxins via the urine, and has long been eaten in soups or taken as a decoction to help ease fluid retention, kidney stones, arthritis, and gout.

Extensive modern research into the medicinal benefits of the cabbage largely confirms its ancient uses in folk medicine. It has been shown to help stimulate the immune system and the production of antibodies, and is a useful remedy for fighting off bacterial and viral infection. It contains sulfur compounds, which may be responsible for its antiseptic, antibiotic, and disinfectant actions, particularly in the respiratory system. It was used as a folk remedy, in the form of soup or tea, for respiratory infections, colds, sinusitis, coughs, and sore throats, and it was an old remedy for tuberculosis.

Cabbage can be used to help heal ulcers. It contains mucilage that coats the lining of the digestive tract and protect it from irritants and excess acid, and an amino acid, methionine, found only in raw cabbage, which promotes healing. It should be taken raw or juiced for best results: 2–3 glasses (12–18 fl oz) of freshly extracted juice taken between meals can help to relieve peptic ulcers, gastritis, heartburn, and ulcerative colitis. Cabbage also benefits the digestion in other ways—it stimulates the appetite and relieves constipation. A traditional Russian folk cure for chronic constipation was half a glass (3 fl oz) of salted cabbage juice taken before each meal. As a tonic to the liver, cabbage has long been used to treat cirrhosis of the liver as well as lethargy, irritability, and headaches, all symptoms associated with a sluggish liver. It is also an old remedy for headaches and hangovers, and was used to dry out alcoholics.

Recent research has confirmed that cabbage contains a substance, glutamine, that can help both peptic ulcers and alcoholism. It has also indicated that cabbage may help to reduce blood sugar and so may be of benefit to diabetics.

Like other brassicas—such as broccoli, cauliflower, and brussels sprouts—cabbage has been thoroughly researched for its protective action against cancer. Cabbage has been found

to help lower the risk of cancer—especially cancer of the colon and the growth of polyps, which are often a prelude to cancer—and, when eaten raw, to help protect against the effects of radiation. The more that is eaten, the better the effect. This is probably due to the many tumor-inhibiting chemical constituents it contains—bioflavonoids, indoles, genistein, and monoterpenes. Cabbage also appears to enhance the body's ability to metabolize estrogen and helps to reduce susceptibility to breast, uterine, and ovarian cancer if eaten regularly.

EXTERNAL USE

Cabbage leaves have a soothing antiseptic and healing effect and they have an ability to draw out toxins from the skin. A cabbage leaf poultice was a traditional remedy for wounds, burns and scalds, boils and carbuncles, bruises and sprains, ulcers, blisters, cold sores and shingles, and bites and stings. Its anti-inflammatory action can benefit swollen and painful joints, arthritis, and gout, and help relieve the pain of neuralgia, sciatica, toothache, headaches, migraine, and lumbago. Traditionally, it was applied over the abdomen and left overnight to treat peptic ulcers and bowel problems. Applied to the lower abdomen during the day and night, it may be useful to soothe cystitis and renal colic, and to relieve fluid retention.

A cabbage poultice applied to the chest and cabbage tea or juice taken internally during the day is reputed to relieve the pain and soreness of a harsh cough and to help to clear a chest infection. If applied to the throat, it can help soothe tonsillitis and laryngitis. The leaves, after being steeped in olive oil, can be applied to chapped skin, chilblains, varicose veins, abscesses, and boils to bring relief.

Cabbage juice can be used as a gargle for sore throats, a lotion for burns, bites, cold sores, acne, impetigo, and squeezed into the ear for earache. Tepid cabbage water is excellent for bathing sore, tired eyes.

Caution: All brassicas should be avoided by people who have overactive thyroid glands. Sauerkraut is high in tyramine and this can sometimes trigger migraine headaches in susceptible people.

HOW TO GROW

By growing different types of cabbage, you can have this vegetable all year around:

Spring cabbage: Sow in mid to late summer, harvest in late spring.

Summer cabbage: Sow in mid spring, harvest in late summer and early autumn.

Winter cabbage: Sow in late spring, harvest from early to late winter.

Cabbages, as do all brassicas, like sun and rich, alkaline, well-drained soil with plenty of organic matter. Sow seeds about ¹/2 in (1.5 cm) deep in trays or a seed bed, leaving 1¹/2 in (3.5 cm) between seeds. Time your sowing according to variety (*see above*). When plants are about 3 in (7.5 cm) high, replant 12–18 in (30–45 cm) apart, in staggered rows 18–24 in (45–60 cm) apart. Harvest cabbages as required, then dig up and dispose of the roots.

KALE *B. oleracea,* Acephala group

Kale was probably the earliest kind of cultivated cabbage, and it is the nearest to the original wild cabbage—the colewort. These early cabbages probably had more stalk than anything else, and kale still has tall, thick stems.

INTERNAL USE

Kale, like cabbage, is a good source of minerals and vitamins, notably beta-carotene, vitamins C and E, iron, folic acid, and a form of calcium that is easily absorbed by the body. It is similar in therapeutic value to its relatives broccoli, cabbage, brussels sprouts, and cauliflower, being rich in antioxidants and other powerful cancer-fighting phytochemicals, including a substance called sulforaphane that blocks the action of several carcinogens. It also contains indoles, which enhance the liver's metabolism of estrogen, and so helps to speed up its excretion from the system, thereby reducing the risk of breast cancer. Indoles and other substances, such as beta-carotene and bioflavonoids, also stimulate the production of those enzymes that protect against cancer.

Kale makes an excellent nutritious tonic for anyone who is anemic or feeling tired, lethargic, and run down. The rich iron content is readily absorbed, due to the presence of vitamin C, and

acts to enhance energy as well as immunity and healing.

Kale is best cooked only lightly, either steamed or stir-fried, to preserve its therapeutic properties. The addition of such spices as cilantro and cumin, or herbs such as rosemary or thyme, may help to neutralize some of this vegetable's gas-producing properties.

HOW TO GROW

Kale should be sown in seed beds in late spring about ¹/2 in (1.5 cm) deep, and thinned to 3 in (7.5 cm) apart when seedlings are large enough to handle. In midsummer, transplant them 18 in (45 cm) apart in soil that was well watered the previous day. Kale needs firm soil with garden compost or well-rotted manure added the previous year. Water in dry weather, earth up and stake in the autumn. To harvest, remove individual leaves from the stem with a knife from mid to late winter. For continued cropping, avoid stripping the plant.

CAULIFLOWER *B. oleracea,* Botrytis group

The cauliflower is thought to have been native to Turkey, Syria, and Egypt, and was apparently mentioned in texts around 540 B.C. It was first recorded in Britain by the herbalist John Gerard, but it was not a frequently eaten vegetable until the end of the 18th century. Although it originated in the East, the cauliflower will survive frosts and light freezes, which many people believe improve its flavor. Mature cauliflowers are more frost-resistant than seedlings.

INTERNAL USE

Cauliflowers are a good source of minerals and vitamins, particularly vitamin C, folic acid, potassium, and bioflavonoids. They are low in calories, which is helpful for dieters, and high in fiber, ensuring healthy bowel function and protecting the bowel from the damage caused by irritants and toxins. Like other brassicas, such as cabbage, broccoli, and brussels sprouts, cauliflower contains substances that are thought to help to reduce the risk of cancer, particularly of the breast and the colon. As do all crucifers, it contains antioxidants, which capture free cals that cause damage to cells and predispose to heart disease, degenerative diseases such as arthritis, and to cancer development. They also neutralize chemicals in the body that activate carcinogens. The crucifers may also help to cleanse the body of carcinogens found in polluted air, pesticides in food, and other environmental sources. The indoles and other anticancer substances present in crucifers have the ability to increase the secretion of glutathione, which can destroy carcinogens and help to enhance the secretion of enzymes that speed up detoxification and, thereby, protect cellular DNA from damage caused by carcinogens. It is important not to cook cauliflower for long, since the indoles are destroyed by heavy cooking.

HOW TO GROW

There are varieties of cauliflower that can be grown for summer, autumn, and winter cropping. Cauliflowers need alkaline soil and are harder to grow than broccoli or sprouts. The soil should be firm, fertile, and well-limed. Do not dig before planting. For varieties that crop in late summer and autumn, sow seeds between mid and late spring about ¹/₂ in (1.5 cm) deep in a seed bed. Transfer to their cropping position at 6–8 weeks, when they have 4 leaves, and plant out in staggered rows 18–24 in (45–60 cm) apart. Mulch and keep plants well watered. Early cauliflowers can be sown mid autumn, overwintered indoors and planted out in spring. To harvest, cut the heads off and dig up and dispose of the root.

BRUSSELS SPROUTS *B. oleracea,* Gemmifera group

As their name indicates, brussels sprouts were first recorded, around 600 years ago, as growing in Belgium. They reached England and France by the 19th century, where they have enjoyed mixed popularity ever since.

INTERNAL USE

Brussels sprouts are a good source of beta-carotene, folic acid, antioxidant vitamins A, C, and E, bioflavonoids, iron, potassium, and fiber. Like their relatives—cabbage, broccoli, and cauliflower—sprouts are also thought to help to

and to regulate the body's estrogen balance. They contain a substance called sulforaphane, which helps to stimulate enzymes to cleanse the body of carcinogens. They also contain indoles which, along with the antioxidants, help to protect against cancer, notably breast cancer, by speeding up the metabolism and removal of estrogen from the body. The folic acid contained in brussels sprouts is beneficial, particularly to pregnant women, since it aids the development of the baby's brain and spinal cord that occurs during the first few weeks of pregnancy. Brussels sprouts are best steamed or lightly cooked, otherwise their indole and vitamin C will be destroyed. Vitamin C is vital for the efficient absorption of iron from food, a healthy immune system, healthy skin and cardiovascular system, and it may help prevent damage caused to the body by free radicals.

HOW TO GROW
Brussels sprouts like firm, fertile soil, and a sunny but sheltered spot. They are usually cultivated in the same way as cabbages (*see page 28*), but they can also be started off in seed trays indoors and planted out, about 24 in (60 cm) apart, when the seedlings have developed 4 or 5 leaves. As the plants grow, earth them up if necessary to prevent them falling over. Individual plants can yield up to 2 lb (1 kg) of sprouts. Harvest them when they are the size of walnuts and still closed. Snap or cut them off the plant, starting from the bottom. Discard the roots after harvesting.

BROCCOLI *B. oleracea,* Italica group
Broccoli, like cabbage, was apparently grown by the ancient Egyptians, Greeks, and Romans and is said to have originated in Crete, Cyprus, or the Eastern Mediterranean. It was highly valued medicinally in early times, and was used to treat headaches, diarrhea, stomach disorders, gout, and even deafness. There are two types of broccoli: calabrese, which is harvested in summer and autumn; and sprouting broccoli, which can be either purple or green and is harvested in winter. Both are delicious when steamed or lightly stir-fried.

INTERNAL USE
Broccoli is rich in nutrients, notably antioxidant vitamins A and C, beta-carotene, calcium, potassium, iron, and folic acid, and has properties very similar to its relatives. Its abundant antioxidants help to protect the body against damage caused by free radicals and so protect against degenerative diseases such as arthritis, heart disease, and cancer. There is now some evidence to suggest that people who eat more broccoli are less likely to develop cancer of the lungs, breast, cervix, colon, prostate, larynx, esophagus, and bladder. The indoles have been shown to speed up the metabolism and removal of estrogen from the body, and so help to prevent breast cancer. The folic acid in broccoli can help to prevent the virus that is related to the development of cervical cancer, to protect the lungs against cancer, and it is vital to pregnant women for the normal development of the brain and spinal cord in the baby.

Broccoli is also a good source of chromium, a substance that helps to regulate insulin and blood sugar. It has been shown to increase the efficiency of insulin so that the body requires less of it—a fact that is useful to non-insulin-dependent diabetics. Broccoli is high in soluble fiber, which not only helps to ensure healthy bowel function, but also helps to reduce the levels of cholesterol in the blood.

Broccoli is best eaten raw or lightly cooked, steamed, or stir-fried. Overcooking and boiling tends to reduce the vitamin C content and destroys the protein and cancer-protecting substances, such as indoles.

HOW TO GROW
Broccoli and calabrese grow best in soil that is firm, not loose, and not too rich in nitrogen. Calabrese is usually sown in the final position. The seedlings should be thinned to 2 in (5 cm) when they are large enough to handle, and kept well watered. Calabrese can be harvested in late summer or early autumn. Broccoli seeds should be sown outdoors in a seed bed in late spring, 1 in (2.5 cm) apart. When the plants are about 4 in (10 cm) high, transplant them 4 in (10 cm) apart in all directions. Keep them well watered and harvest late spring or early summer.

Brassica
recipes & remedies

Spices added when serving or cooking—such as coriander seed, cumin, caraway, or tarragon—will help to reduce flatulence. Cabbage is most effective when eaten raw as some antioxidant, anticancer, and estrogen-balancing compounds are destroyed by cooking.

Salted cabbage juice

Excellent for constipation and a sluggish liver.

2 lb (1 kg) cabbage, finely shredded
Salt to cover

Place the shredded cabbage in a large dish. Pour salt over it and let stand for half an hour until the cabbage is moist. Strain the juice and take 1–2 teaspoons (5–10 ml) 3 times daily. Store in the refrigerator.

Cabbage syrup

A soothing remedy for colds, coughs, bronchitis, and sore throats.

6 oz (150 g) cabbage leaves, finely
shredded
Honey to cover

Cover the shredded leaves with honey. Let stand overnight and then press through a sieve to collect the syrup. Take 1 teaspoon (5 ml) every 2 hours for as long as the symptoms persist. Store in the refrigerator.

Cabbage leaf poultice

Apply to the appropriate area to ease the pain of arthritis, cystitis, coughs or sore throats.

Use as many green leaves as possible, cutting out the middle ribs. Warm them in a little hot water, iron them (with a dish towel over the leaves), or hang them over a radiator until dry. Crush the dry leaves with a rolling pin and apply several layers, holding them in place with a bandage. Change the leaves every few hours.

Cauliflower with dill vinaigrette

The excellent digestive and warming properties of dill are beneficially combined here with cauliflower, which, like other brassicas, can be hard to digest.

1 medium-sized cauliflower
1 tablespoon Dijon mustard
1 tablespoon cider vinegar
3 tablespoons extra virgin olive oil
Chopped fresh herbs: 2 tablespoons
        dill and 1 tablespoon parsley
Salt and freshly ground pepper

Cut the cauliflower into bite-sized pieces. Steam for 2–3 minutes, or until just tender. Refresh under cold water and drain well. Place in a salad bowl. Put the mustard and cider vinegar into a small bowl and whisk to combine. Slowly whisk in the olive oil. Add the herbs and season with salt and pepper. Pour the dressing over the cauliflower. Toss lightly, then chill before serving.

Turnip *Brassica rapa,* Rapifera group

The turnip, like the rutabaga, is a native of Europe and a member of the Crucifereae, or cabbage, family. The turnip has been cultivated since about 3000 B.C., probably first in Mesopotamia. It was popular with the ancient Greeks, and was introduced into other parts of Europe by the Romans. Turnips were grown in the kitchen gardens of the medieval monasteries and they were eaten fresh or preserved in vinegar or brine. They were considered beneficial to the stomach, to moisten "dry intestines," and as a diuretic.

TURNIPS CAN HELP TREAT

- *Acne*
- *Arthritis*
- *Bladder infections*
- *Boils and abscesses*
- *Bowel disorders*
- *Congestion, coughs, and colds*
- *Chilblains*
- *Constipation*
- *Eczema*
- *Fluid retention*
- *Gout*
- *Low immunity*
- *Vitamin and mineral deficiency*

INTERNAL USE

Turnips are highly nutritious vegetables, containing the vitamins A and C, and minerals including calcium, phosphorus, magnesium, sulfur, iodine, and potassium. Turnip tops are particularly high in vitamins A and C, calcium, iron, and copper. As a result, turnips have gained a reputation as an energy-giving tonic and as a remedy to cleanse the blood, useful for clearing skin problems. Turnip greens, with their high calcium content, are good for building and maintaining healthy bones and teeth, and they are good for both children and post-menopausal women. The whole plant was traditionally used to combat scurvy and for nutritional deficiencies responsible for lethargy and low spirits.

Turnips have long been valued for their beneficial effect on the urinary system. They have been used as a remedy for fluid retention, urinary infections, obesity, gout and arthritis, and kidney stones (largely formed of uric acid).

The sulfur compounds found in turnips contribute to their valuable antibacterial properties, which are particularly beneficial to the respiratory system. Turnip juice can be an effective decongestant; 1 teaspoonful (5 ml) of juice taken three times daily makes a useful children's remedy for colds, coughs, and congestion.

The fiber in turnips also benefits the digestive system, ensuring normal bowel function and helping to prevent constipation and other bowel problems, which may help reduce the risk of such diseases as bowel cancer.

Recent research has indicated that turnips may enhance general immunity. Both the root and the green tops are high in substances called glucosinolates, which have been reported to help to block the development of cancer. Raw turnips are higher in glucosinolates than cooked ones. The dark green leaves are also rich in chlorophyll and in carotenoids, including beta-carotene, which research has indicated to be anticarcinogenic. They are delicious when cooked, steamed until tender. Turnips have been shown to accelerate the metabolism of estrogen, which may help to guard against the development of estrogen-dependent tumors, such as breast cancer.

EXTERNAL USE

Turnips are reputed to have a soothing and healing effect on the skin and they have been used in hot poultices in order to draw out boils and abscesses and to help to heal chilblains. They can also be applied in poultices to ease aching muscles and painful joints associated with rheumatism, arthritis, and gout.

HOW TO GROW

Grow in light shade in alkaline soil, which should be prepared with plenty of garden compost or well-rotted manure dug in the autumn before planting. Sow seeds outside every three weeks from late spring to midsummer for a continuous supply, and again in autumn for winter eating. Seeds should be sown 1/2 in (1.5 cm) deep, in rows 1 ft (30 cm) apart. Thin to 4 in (10 cm) apart when large enough to handle, and then to 6 in (15 cm) 3 weeks later. They should be kept watered, especially in dry weather.

For a good flavor and texture, turnips should be harvested when young, after about 70 days.

Turnip
recipes & remedies

Turnip leaves can be cut to about ½ in (1.5 cm) above the roots within 4 weeks of sowing, and then steamed or lightly boiled.

Turnip syrup

Makes an effective decongestant.

1 turnip, sliced
Honey to cover

Cut a turnip into 3 or 4 slices and cover each side with honey. After 2–3 hours the honey will have drawn out the juice to form a syrup. Take 1–2 teaspoons every 2–3 hours while symptoms persist. Store in the refrigerator.

Turnip purée

A traditional remedy for bronchitis.

1 lb (450 g) turnips, sliced
3 teaspoons milk or olive oil

Steam or boil the turnips in a little water. Once soft, purée in a blender, adding a little milk or olive oil. Take 2–3 tablespoons 2–3 times a day.

Turnip poultice

Apply hot to draw out boils and abscesses, to heal chilblains, and relieve painful joints.

¹/2 quantity turnip purée (see previous recipe)
2 pieces gauze

Light cotton bandage
Place sufficient of the turnip purée to cover the affected area between 2 pieces of gauze. Bind it to the affected area with the bandage (*see page 161 for further information*).

Turnip and dill soup

A highly nutritious soup, rich in vitamin A, calcium, and magnesium, which cleanses and strengthens the immune system and is excellent for a weak digestion.

1 tablespoon olive oil
1 large onion, finely chopped
4 cups (450 g) peeled and finely
      chopped turnips
3 tablespoons fresh dill
Salt and freshly ground pepper
3³/4 cups (850 ml) water or
      vegetable stock

Heat the oil in a large saucepan over low heat. Add the onion, cover, and cook for 10 minutes, or until translucent but not browned. Add the turnip and three quarters of the dill, stir well, and season to taste with salt and pepper. Cover and cook gently for 30 minutes, adding a little of the water or stock if it gets too dry. Add the remaining water or stock, increase the heat, and bring almost to a boil. Remove the pan from the heat and allow to cool a little. Purée the soup in a food processor or blender until smooth. Return to the rinsed-out saucepan and reheat. Garnish with the remaining dill and serve.

Pepper _Capsicum annuum_ var. _annuum_

There are many varieties of pepper: sweet peppers, bell peppers, chili peppers, paprika, cayenne, tabasco. Some are sweet and mild, others hot and pungent. They all derive from the same wild species (_C. annuum_) that came originally from Central and South America: in fact, peppers were grown in Mexico as far back as 7000 B.C. Pre-Columbian ceramics decorated with peppers confirm that the Aztecs cultivated and used them, and by the time peppers were introduced to Europe by Christopher Columbus in 1493, most of the varieties with which we are now familiar had been developed.

SWEET PEPPERS
CAN HELP
TREAT

- _Allergies_
- _Cardiovascular problems_
- _Respiratory infections_

HOT PEPPERS
CAN HELP
TREAT

- _Congestion, coughs, and colds_
- _Chilblains_
- _Fevers_
- _Gastrointestinal infections_
- _Inflammation_
- _Neuralgia_
- _Period pain_
- _Poor circulation_
- _Sinusitis_

INTERNAL USE

Peppers are rich in beta-carotene and vitamin C, natural antioxidants that help to protect against degenerative diseases, cancer, and cardiovascular diseases such as atherosclerosis and angina. Peppers not only contain bioflavonoids, which research has indicated have anticancer properties, but also phenolic acids and plant sterols, both of which may help inhibit the formation of tumors. Regularly eating fresh, raw peppers enhances immunity and helps the body to fight off infections such as colds, flu, and coughs, and may help protect against allergies such as eczema and asthma.

Cayenne pepper and other hot, spicy peppers are powerful stimulants, particularly to the heart and circulation, and make an excellent warming remedy for those with poor circulation and associated problems such as chilblains, cold extremities, tiredness, and depression. If eaten or taken in a hot drink at the onset of a cold or flu, cayenne increases sweating and so enhances the body's fight against infection. Cayenne has a bactericidal action, it is rich in vitamin C, and it makes a wonderful remedy for the respiratory system. The pungency of cayenne acts as an effective decongestant in the chest and upper respiratory tract, easing expectoration and relieving stuffiness, congestion, and sinusitis.

Cayenne also has a revitalizing effect on both body and mind, dispelling tiredness, lethargy, nervous debility, and depression. The burning sensation experienced on the tongue caused by eating cayenne or chilies sets off messages to the brain to stimulate the secretion of endorphins, which are opiate-like substances that can block pain and induce a feeling of well-being, even euphoria. Research has shown that hot peppers have an analgesic effect and can ease the pain of toothache, shingles, and migraine.

Cayenne's pungency has a stimulating effect throughout the digestive tract, improving appetite, digestion, and the absorption of food. It can help to relieve symptoms of a weak digestion, such as diarrhea, gas, nausea, and pain.

In the reproductive system, cayenne's warming properties help to relieve pain caused by poor circulation and can bring on delayed periods. According to modern research, cayenne can help ease circulation problems. It helps to prevent blood clots and lower harmful cholesterol.

EXTERNAL USE

Hot peppers can be used as local stimulants in ointments and liniments to relieve arthritic and muscular pain, neuralgia, bruises, and back pain. They are an excellent remedy for chilblains (if the skin is unbroken). Their pungency helps to bring out inflammation and, by numbing the skin, to relieve pain. Cayenne pepper powder placed in woolen socks makes an excellent remedy for poor circulation and chilblains.

HOW TO GROW

Peppers like moist, free-draining soil with plenty of organic matter and a warm, sheltered spot. Sow seeds indoors under glass in spring, and transplant to a pot when they have grown 3–4 leaves, and keep warm. Continue to grow in a pot in the greenhouse, or, if they are intended for outdoors, harden them off and plant them outside in early summer, 18 in (45 cm) apart. They should be secured to stakes, and will need humidity in order to set. Peppers can either be cut when green, in midsummer if in a greenhouse, or in late summer/early autumn if grown outdoors. If yellow or red peppers are preferred, they can be left on the plant to ripen.

Pepper
recipes & remedies

If the pungency of hot peppers proves hard to swallow, it is advisable to start with small amounts and gradually build up a tolerance. It is best avoided by those prone to overheating and acidity of the stomach as it may aggravate the problem.

Cayenne, elderflower, and peppermint tea

A remedy for fevers, flu, colds, and congestion.

2 cups (570 ml) boiling water
¹/₂ oz (15 g) elderflowers
¹/₂ oz (15 g) peppermint leaves
1 pinch cayenne powder

Pour the water over the herbs and let them infuse for 15 minutes. Drink 1 cupful, hot, 3–6 times daily while symptoms persist.

Heating liniment

For muscle pain, rheumatism, arthritis, neuralgia, sprains, and strains. The oils below can be bought at health and beauty shops.

¹/₄ teaspoon (1.25 ml) capsicum tincture (*see page 159*)
25 drops rosemary oil
25 drops lavender oil
2 fl oz (50 ml) almond oil

Combine the ingredients together and massage daily into the affected area.

Gargle or mouthwash

To combat throat and mouth infections.

1¹/₂ oz (40 g) sage leaves
1¹/₂ oz (40 g) thyme
2 teaspoons (10 ml) cayenne pepper
1¾ cups (500 ml) cider vinegar

Combine the ingredients together in a large jar and leave them covered to macerate for 2 weeks. Strain and store in an airtight sterilized bottle. Use 1 teaspoon (5 ml) in a little warm water 2–3 times daily.

Chicory *Cichorium intybus*

Chicory can often be seen growing wild on embankments and roadsides, with leaves like its relative, the dandelion, and exquisite bright blue, daisy-like flowers. The many cultivated varieties such as radicchio (*see photo right*) or Belgian endive (*see below*) maintain the medicinal benefits of the wild plant, and the roots, leaves, and flowers can all be used (although roots have the strongest effect). The leaves of the different varieties add a pleasantly bitter taste to salads, while the flowers add interest and color.

CHICORY CAN HELP TREAT

- Arthritis
- Constipation
- Fluid retention
- Gout
- Headaches
- Indigestion and heartburn
- Liver and gall bladder problems
- Tiredness and lethargy
- Urinary infections

INTERNAL USE

Chicory has been well known as a vegetable and as a medicine since the time of the ancient Egyptians; it is mentioned on a papyrus dating back about 4,000 years. The Greeks and Romans enjoyed it as a vegetable and the Roman physician Galen referred to chicory coffee as "the friend of the liver," recognizing its benefit to the liver and gall bladder (for which it is still valued today). In the Bible, it was one of the bitter herbs that God commanded the Israelites to eat with lamb at Passover.

Like the dandelion, chicory stimulates the flow of saliva and other digestive juices. It enhances the appetite, promotes digestion and absorption, and can be used to improve a sluggish digestion and to relieve indigestion and heartburn. Chicory can help to stimulate the function of the bowels, the liver, and gall bladder. It makes an effective remedy for mild constipation and for conditions that are often associated with a sluggish liver, such as skin problems, headaches, lethargy, and irritability. It may be helpful in treating gallstones.

Chicory also has a diuretic effect, enhancing the elimination of fluid and toxins from the system. This helps to cleanse the blood and can be helpful to people suffering from arthritis and gout. Its antibacterial properties combined with the diuretic action may help to relieve urinary infections such as cystitis and urethritis.

HOW TO GROW

Chicory likes a rich soil, with plenty of garden compost or well-rotted manure. The seeds should be sown outdoors in late spring or early summer, and thinned to 9 in (23 cm) apart when seedlings are large enough to handle. They should be kept well watered, especially in dry weather. To harvest chicory, cut the leaves about 1 in (2.5 cm) from the base and lift the root out of the soil using a fork. The roots can then be stored until required for making remedies or coffee (*see opposite*), or forced for blanching (*see below for instructions*). Chicory root can also be dried by chopping it and placing it on a tray in a low oven and roasting until it is dry and brittle.

To blanch chicory (Belgian endive)
Roots should be cut down to 9 in (23 cm) before being stored in a cool place until required for forcing (early to late winter).

Plant 3 or 4 roots in a 10 in (25 cm) diameter pot and water thoroughly. After planting, cover with an upended pot to block out light, and keep it in a warm place.

The blanched leaves should be ready for cutting about a month from planting. Once the leaves have been cut, the roots can be composted or used for remedies (see opposite).

Chicory
recipes & remedies

Varieties of chicory with a long taproot are dried, roasted, and ground to blend with coffee.

Chicory root decoction

Used to help improve digestion.

2 oz (55 g) fresh chicory root or
    1 oz (25 g) dried root, washed and
    chopped
2¹/2 cups (700 ml) water

Add the washed and chopped chicory root to the water in a pan and bring it to the boil. Simmer for 20 minutes. Strain and drink 1 cupful 3 times daily.

Chicory coffee

Chicory helps to counteract the stimulating effects of caffeine, and acts as an excellent digestive.

2 oz (55 g) fresh chicory root or
    1 oz (25 g) dried root.

Place root on baking tray in oven at 350°F (180°C) for around one hour, or until brittle. Grind the chicory root in a coffee grinder and use 1–2 teaspoons (5–10 ml) of the powder per cup of hot water.

Cucumber *Cucumis sativus*

The cucumber is a member of the Cucubitaceae, or gourd, family, which includes melons, pumpkins, zucchinis, and other squashes. It comes originally from the East, where it has been grown for thousands of years and where its cooling, refreshing, and thirst-quenching properties were appreciated in the heat. Those people from more temperate climates, however, viewed the cucumber very differently: the English herbalist John Gerard claimed that it "filleth the veines with naughty cold humours."

CUCUMBERS CAN HELP TREAT

- *Arthritis*
- *Bladder infections*
- *Eczema*
- *Fevers*
- *Fluid retention*
- *Gastritis*
- *Gout*
- *Heat rash*
- *Inflammatory eye problems*
- *Insect bites and stings*
- *Overheating*
- *Sunburn*
- *Urticaria*

INTERNAL USE

Cucumber seeds are mentioned in 18th-century medical pharmacopias as being one of the four coldest seeds, useful for cooling hot, inflammatory problems, and cucumbers are still popular as a refreshing summer food today. In spite of consisting of 96.4% water, they are, when the peel is left on, nourishing. They contain vitamins A and C, and minerals including sulfur, manganese, phosphorus, silicon, sodium, calcium, and potassium. This mineral content helps to prevent nails from splitting, and to maintain healthy hair; the potassium helps to regulate blood pressure.

The low calorific value of cucumbers makes them popular with dieters. Their mild diuretic action may help weight loss where there is fluid retention, and is helpful for relieving bladder infections. Cucumbers have earned a reputation as a cleansing remedy for increasing the elimination of wastes, including excess uric acid—thus helping those suffering from arthritis and gout—in fact they have long been used as a remedy for inflammatory joint problems.

The cooling properties of the cucumber have been used in many different ways. Cucumber was used as a folk remedy for fevers—it was given as a juice or as cucumber water, or even placed alongside a sick infant, when the heat of the fever was said to be absorbed by the cucumber. It has also been used to remedy excessive heat and inflammation in the body. Today it is still valued for its cooling properties, and in France, cooked cucumber is a popular remedy to aid liver function, and for treating intestinal disorders and infections.

EXTERNAL USE

Cucumbers have long been famous as a cooling and soothing remedy for problems of the skin. The juice can be used to soothe urticaria and eczema, as well as prickly heat and sunburn, and makes a noticeable improvement fairly quickly. Cucumber juice mixed with equal parts of rosewater can be applied to skin problems, as well as to chapped lips, and helps to reduce the pain and inflammation of insect bites and stings. Many people apply cucumber slices to their eyelids to cool sore or inflamed eyes and to tone up the surrounding skin. Cucumber is often used to cleanse and tone the skin, particularly when it is oily and prone to pimples or blemishes, and to soften hard skin.

HOW TO GROW

There are different types of cucumber: outdoor (ridge) cucumbers and greenhouse cucumbers.

Outdoor varieties should be grown on ridges of soil in a sunny, sheltered position and their shoots allowed to trail on the ground. Dig holes to one spade's depth, 2 ft (60 cm) apart and fill them with a mixture of manure and soil. Sow seeds in late spring (earlier if under cloches) and water well. Cucumbers do not transplant well and are best planted in their final position. After 5–6 leaves appear, pinch out the growing points. Cucumbers should be harvested when fully ripe (early autumn).

Greenhouse varieties need warmth, humidity, and plenty of watering and feeding. Sow two seeds about 1/2 in (1.5 cm) deep in a 3 in (7.5 cm) pot in late spring. When 2–3 leaves have developed, thin to one plant per pot. Support growing plants with canes or horizontal wires. Pinch out the tips of the leading shoots, feed every 2 weeks with potash fertilizer once the fruits have started to grow, and mist regularly with warm water. Greenhouse cucumbers can be harvested from mid to late summer.

Cucumber
recipes & remedies

As part of a meal, especially one that contains hot and spicy dishes, cucumber-based dishes such as Indian raita can help to cool the stomach and prevent irritation.

Cucumber juice

Can be used to help bring down a fever, especially in children.

1 cucumber, peeled and thinly sliced

Place the cucumber slices in a bowl for 2 hours and then collect the juice by filtering it through fine cheesecloth, pressing the slices thoroughly to squeeze out all the liquid. Take 1–2 teaspoonfuls every 2 hours.

Grated cucumber

A useful remedy to soothe sunburn, stings, and other inflammatory skin problems.

1 cucumber, peeled and grated

Grate cucumber so that it is semi-liquid, and massage it into the affected areas.

Cucumber raita

A cooling accompaniment to a hot spicy meal. The mint aids digestion and the natural yogurt benefits the bacterial population in the intestines.

1 cucumber, peeled and diced
1/2 cup + 2 tablespoons (150ml) thick yogurt
2 tablespoons chopped fresh mint leaves
Fine sea salt
A squeeze of lemon juice
A sprig of fresh mint, to garnish

Combine the diced cucumber, yogurt, and chopped mint in a bowl. Add sea salt and a squeeze of lemon juice to taste. If you like, transfer to a serving dish. Chill, then garnish with a sprig of fresh mint and serve.

Cucumber and coriander salad

A cooling salad, which cleanses the system of harmful toxins—excellent for hot summer days.

1/4 head lettuce, shredded
1 bunch watercress, stemmed
1 cucumber, peeled and thinly sliced
A few sprigs of fresh coriander, chopped
Extra virgin olive oil
Fresh lemon juice
Fine sea salt

Arrange the lettuce on a platter or in salad bowl. Add the watercress and cucumber, and sprinkle with the chopped coriander. Drizzle a little olive oil over the salad, add a squeeze of lemon juice and a little sea salt to taste, toss lightly, and serve at once.

Squash, pumpkin, zucchini, and marrow _Cucurbita_ spp.

Squashes, pumpkins, and zucchini (collectively known as cucurbits) are all members of the Cucurbitaceae family, and are some of the most ancient vegetables in existence. The cucurbits also include some of the most curious-looking ones, and many are named according to their shapes: turban gourds, crookneck squash, banana squash. They also include some of the biggest vegetables—the largest recorded pumpkins have weighed in at more than 1,000 lb. (500 kg.)

CUCURBITS CAN HELP TREAT

- _Bowel disorders_
- _Burns_
- _Colitis_
- _Gastritis_
- _Headaches_
- _Indigestion_
- _Overheating_
- _Peptic ulcers_
- _Prostate problems_
- _Worms_

INTERNAL USE

Cucurbits have been used medicinally for centuries, but it was not until the beginning of the 19th century that one of their most valuable therapeutic properties was discovered—the ability of pumpkin seeds to help rid the body of worms. This is due to the presence of a substance known as cucurbitive, which effectively treats roundworms, threadworms, and tapeworms without irritating the bowel.

All cucurbits are a rich source of nutrients, notably the natural antioxidants beta-carotene, folic acid, vitamins C and E, which have been shown to help prevent cancer, and minerals including potassium, iron, calcium, magnesium, phosphate, copper, and zinc.

Research has discovered traces of other cancer-preventing substances in squash seeds, known as protease trypsin inhibitors, which can prevent activation of viruses and carcinogens in the digestive tract. It is thought that the deep orange squashes in particular, those highest in beta-carotene, can help lower the risk of lung, stomach, esophageal, bladder, prostate, and laryngeal cancers when they are eaten on a regular basis. By deactivating carcinogens, squashes appear to protect against lung cancer in smokers as well as passive smokers.

Eaten as a vegetable, the fiber in squashes has a beneficial effect in the bowel, ensuring regularity and protecting against such disease of the bowel as diverticulitis and cancer. The fiber in squashes binds to toxins and carcinogens and helps evacuate them from the body.

Pumpkin seeds are rich in fiber, protein, essential fatty acids, vitamins B and E, as well as zinc, iron, and calcium. Not only do they enhance immunity but they have particular significance for the male reproductive tract, where they may help to reduce benign enlargement of the prostate gland. Research has now shown that the amino acids alamine, glycine, and glutamic acid in the seeds can reduce the symptoms of prostate enlargement, such as the frequency of urination that can lead to disturbed sleep. The high zinc content of pumpkin seeds also helps to balance male hormone levels, and so may help to prevent prostate problems.

EXTERNAL USE

The pulp of raw pumpkin can be applied as a poultice to soothe burns and headaches, and the seeds can also be pounded in oatmeal and applied to the skin to remove blemishes.

HOW TO GROW

Cucurbits like warm, moist conditions and need rich soil. They prefer partial shade and they need to be grown well away from other crops so that they do not smother them. Seeds should be sown under cloches in late spring, or in the open in early summer. Plant seeds sideways 1 in (2.5 cm) deep, in 3 in (7.5 cm) diameter pots. Seedlings can be planted out once they have grown 4–6 leaves. Put in firm supports if required, and plant in compost-filled holes. Water the roots, not the leaves, especially in dry weather. Marrows should be planted 3–4 ft (1–1.2 m) apart each way, pumpkins 4 ft (1.2 m) apart, and zucchinis 3 ft (1 m) apart with 3 ft (90 cm) between rows. Harvest cucurbits with their stalks on, from midsummer onward.

Cucurbit
recipes & remedies

As well as being eaten as a main course or side dish, cucurbits such as cucumbers and pumpkins can be made into jams and pickles.

Pumpkin seed and honey paste

A remedy for expelling worms.

1¹/4–1¹/2 oz (35–40 g) pumpkin seeds, hulled
1¹/4–1¹/2 oz (35–40 g) honey

Pound the pumpkin seeds with a pestle and mortar to make a paste. Mix the paste with equal parts of honey and take first thing in the morning before breakfast in 3 doses, 20 minutes apart. Continue for 2 more days.

Pumpkin and oatmeal paste

An excellent remedy for skin blemishes.

2 oz (50 g) pumpkin seeds, hulled
2 oz (50 g) oatmeal

Pound the pumpkin seeds to make a paste. Combine this with the oatmeal and apply to the skin for 10–15 minutes. Rinse off with warm water or rosewater.

Pumpkin seed tea

Helpful for prostate disorders.

3¹/2 oz (100 g) pumpkin seeds (with shells)
4 cups (1 liter) water

Simmer the pumpkin seeds in water for 20 minutes, strain and take a glassful (6 fl oz) 3 times a day.

Steamed zucchinis

A nutritious remedy to help regulate the bowels.

1 lb (450 g) zucchinis, chopped

Chop the zucchinis into 1 in (2.5 cm) lengths. Place them in a steamer and steam for 5–10 minutes, or until tender.

Artichoke *Cynara scolymus*

The artichoke is a magnificent architectural plant for the back of an ornamental border or herb garden. It is a perennial in temperate climates and has large silvery leaves and gray-blue, thistle-like flowers. It is one of the oldest cultivated vegetables, and was grown by the ancient Egyptians, Greeks, and Romans alike. It was introduced to Europe by the Arabs in the 15th century and to Britain in the 16th century. The Arabs recommended the leaves as a medicine, particularly to treat the liver and sluggish digestion.

ARTICHOKES CAN HELP TREAT

- *Acne*
- *Arteriosclerosis*
- *Arthritis*
- *Atherosclerosis*
- *Eczema*
- *Fluid retention*
- *Gout*
- *Heartburn*
- *Indigestion*
- *Nausea*
- *Poor appetite*
- *Urticaria*

INTERNAL USE

Europeans have long respected the artichoke as a "friend of the liver." The bitters in artichoke leaves act to stimulate the flow of digestive juices and aid bile secretion from the liver and gall bladder. This explains why extracts of artichoke have traditionally been included in bitter alcoholic aperitifs and *digestifs*, to whet the appetite before a meal and ease the digestion of food afterward, as well as to support a liver that has been overworked by the excesses of a rich and heavy meal with alcohol.

Modern practice and research support the ancient use of artichoke as a medicine. In Europe particularly, the artichoke is popular as a medicine to lower cholesterol and triglycerides, and to treat atherosclerosis and arteriosclerosis. Cynarin, a substance, found in the leaves, has been shown to help improve liver and gall bladder function as well as to lower cholesterol levels. Artichoke leaves in teas and tinctures are used by modern herbalists to help remedy weak digestion, poor appetite, heartburn, nausea, liver insufficiency, and skin problems such as acne, eczema, and urticaria. Artichokes also have diuretic properties, enhancing the elimination of fluid and toxins from the system. So, with their beneficial action on the liver as well as on the kidneys, artichokes make a good cleansing remedy, helping to clear the skin and to relieve arthritis and gout.

HOW TO GROW

Artichokes grow 3–5 ft (1–1.75 m) tall and like rich, light, well-drained soil. They will not grow well in heavy clay. They prefer a warm climate, and may not survive the winter unless they are grown in the milder parts of Europe and North America. In any event, they should be protected from frost.

Artichokes are usually grown from offsets collected from healthy productive plants (*see below*). In warm climates where artichokes grow as perennials, they are more productive in their second and third years and do not normally continue to grow after 5 years. Artichoke shoots should be planted shallowly in mid to late spring, 4 ft (1.2 m) apart, and mulched and watered well. The heads should be harvested when the leaves are tightly wrapped and still green from the second year onwards in midsummer. If the artichoke heads are left on the plant, they will turn into huge, beautiful thistle flowers that will brighten your garden.

The stems should be cut back in late autumn, and they should be protected from winter frost by earthing up.

To propagate, select healthy shoots from existing plants that are more than 3 years old and remove, using a spade or large, sharp knife, making sure that each shoot retains some of its roots.

Remove the buds as soon as they appear in the first year. This will encourage further growth. The heads can be harvested from the second year onward.

Artichoke
recipes & remedies

Artichokes can be steamed or cooked in boiling water for 15–30 minutes, drained, and left to cool. The tender part of each leaf can be eaten. One by one each leaf is torn away from the base and dipped in vinaigrette, butter, or mustard sauce.

Artichoke leaf infusion

To aid liver and gall bladder function and to help lower blood cholesterol.

1 oz (25 g) dried artichoke leaves or
    2 oz (50 g) fresh leaves,
    chopped
2 cups (570ml) boiling water

Place the chopped leaves in a large teapot. Pour on boiling water. Cover and allow to infuse for 10–15 minutes. Drink 1 cupful 3 times daily.

Carrot *Daucus carota*

The humble carrot, a native of Afghanistan, was well known to the ancients. It was discussed by the Greeks in writings dating back to 500 B.C. and was used by Hippocrates in 430 B.C. The familiar garden carrot, now grown all over the world, is the cultivated variety of the wild carrot, an umbelliferous plant, also called Queen Anne's Lace. The name *Daucus* comes from the Greek *daio* meaning to burn, on account of the pungent and stimulating qualities of carrots, particularly the seeds.

CARROTS CAN
HELP TREAT

- *Anemia*
- *Arthritis*
- *Boils and abscesses*
- *Bronchial congestion*
- *Constipation*
- *Cuts and abrasions*
- *Cystitis*
- *Diarrhea*
- *Flatulence*
- *Fluid retention*
- *Gout*
- *Heart and arterial disease*
- *Intestinal infections*
- *Liver problems*
- *Minor burns and scalds*
- *Poor vision*
- *Respiratory infections*
- *Vitamin and mineral deficiency*

INTERNAL USE

Carrots are highly nutritious, being rich in vitamins A, B complex, and C, in minerals including iron, calcium, potassium, and sodium, and in beta-carotene, asparagin, and daucarine. They have long been praised as a restorative remedy, promoting growth and vitality, helping to build healthy tissue and skin, for use in debility, convalescence, mineral deficiency, rickets, dental caries, and anemia. Recently, carrots have been shown to increase hemoglobin and red blood cell counts. Their high vitamin A content has meant that carrots have long been considered excellent for promoting good night vision and for general care of the eyes.

Carrots are renowned for their digestive properties. They regulate intestinal activity, promoting bowel function, and can be useful both in easing constipation and preventing diarrhea. They have the ability to soothe the mucous membranes throughout the digestive tract, which helps to reduce irritation and inflammation. Pureed carrot can be given even to small infants to treat digestive problems. Carrots can be used to relieve flatulence, irritable bowel syndrome, and intestinal infections. A carrot-juice fast for 1 to 2 days is a well-known cleansing therapy for the liver, and can also help to clear up skin problems. An infusion of carrot tops has been used to treat eczema and acne.

In 1960, Russian scientists identified a chemical ingredient in carrots called daucarine, which has been shown to dilate blood vessels, particularly those in the head, helping to protect against arterial and heart disease. Since then, the carotenoids in carrots have been found to have antioxidant properties, which confirms their folk use as a circulatory remedy, since antioxidants help to reduce damage caused by free radicals and thereby to reduce degenerative disease, notably in the heart and arteries.

Beta-carotene is now thought to inhibit the development and growth of tumors, particularly in smoking-related cancers, in the lungs and pancreas. Studies have shown that eating at least one raw or lightly cooked carrot daily may be enough to have this effect.

Caution: Although carrots are very nutritious, eating too many can give the skin a yellow tinge known as carotenemia, which will disappear when consumption is reduced.

EXTERNAL USE

Grated raw carrot can be used in the form of a poultice as an antiseptic and to speed up the healing of wounds, burns, boils, abscesses, and styes. Carrot broth can be applied to chilblains and chapped skin, to soothe itching in eczema, and to treat impetigo and cold sores. It can also be used as an antiseptic mouthwash and as a gargle for sore throats.

HOW TO GROW

Early carrots, for summer eating, are usually short and fat, and main or late crop carrots are longer and more suitable for storing. Both prefer a light, well-drained, sandy soil with plenty of organic matter. Warm the soil for early sowings by using cloches, and rake the soil to a fine tilth before sowing. Sow seed about 1/2 in (1.5 cm) deep, 8 in (20 cm) apart, in late spring or late summer for earlies, and midspring to early summer for later ones. Both should be thinned to 1 in (2.5 cm) apart, with 3–4 in (7.5–10 cm) between rows, when large enough to handle. Early varieties can be harvested as required in midsummer and early to mid winter, and later ones from midsummer to early autumn.

Carrot
recipes & remedies

For maximum nutritional benefit drink carrot juice or eat carrots raw, or lightly steamed or stir-fried. Once cooked, their soothing and anti-inflammatory properties come into their own.

Carrot juice

A remedy for expelling threadworms in children. Taken with honey and a little water, it is also useful for colds and coughs.

3–4 carrots, washed and cleaned

Juice the carrots in an extractor. Adults can take 1 glassful (6 fl oz) daily before breakfast. For children under 12 years, dilute the carrot juice with an equal amount of water.

Carrot broth

Soothes chilblains, cold sores, and impetigo, and can also be used as a gargle for sore throats.

1 lb (450 g) carrots, washed and cleaned
3 cups (850 ml) boiling water

Place the carrots in the water, cook until they are soft, and then blend.

Cooked carrot purée

A remedy for diarrhea in infants. If symptoms persist, consult your doctor.

1 lb (450 g) carrots, washed and cleaned
4 cups (1.2 liters) boiling water
1 teaspoon (5 ml) olive oil
Salt and pepper to taste

Cook the carrots until soft. Blend, strain. Then add a little olive oil and salt and pepper to taste, with sufficient of the cooking water to make up 4 cups (1.2 liters) again.

Arugula *Eruca vesicaria* subsp. *sativa*

Arugula, with its pungent, spicy leaves, is becoming increasingly popular as a salad vegetable. Arugula has been grown at least since the time of the ancient Greeks and Romans and was known for its medicinal values in those times: the Greek physician Dioscorides knew arugula as "a digestive and good for ye belly." It is a native of the Mediterranean region and of eastern Asia, and derives its Latin name from the ancients' observation of the plant: *Eruca* means "downy stemmed" and *vesicaria* means "like a bladder"—a reference to the seed pods. It now grows wild in many parts of Europe and Asia.

ARUGULA CAN HELP TREAT

- *Anemia*
- *Bronchial congestion*
- *Bruises*
- *Congestion, coughs, and colds*
- *Constipation*
- *Inflammation*
- *Poor circulation*
- *Tiredness and lethargy*
- *Weak digestion*

INTERNAL USE

The strong mustardlike flavor of arugula leaves was popular in Elizabethan England both as a food and as a medicine. It has a stimulating effect on the circulation that increases energy and a sense of well-being. Its pungency stimulates the appetite and may improve the digestion and absorption of food. It also helps cleanse the digestive tract, removing stagnant food by its mild laxative action, and its cleansing effect also enhances general health and vitality.

The stimulating properties of arugula leaves can also be felt throughout the respiratory tract, where they aid in clearing congestion by helping to loosen phlegm and easing expectoration from the chest and throat. The seed pods are edible, too, and like the leaves they were long considered to have certain aphrodisiac properties. Dioscorides was aware of this, writing in the 1st century A.D.: "this being eaten raw in any great quantitie doth provoke venery and the seed of it also doth work ye like effect."

The leaves are rich in vitamins A and C and minerals, notably iron, calcium, and potassium. They were once eaten to prevent scurvy, and in large amounts were given as an emetic to induce vomiting to clear toxins from the stomach. The natural antioxidants contained in the leaves enhance immunity and help prevent damage to the body caused by free radicals. Thus, arugula, like other members of the crucifer family, helps to protect against cardiovascular disease, degenerative diseases, and cancer. Arugula leaves are best picked when they are young and tender to add to salads, and they go better with bland-tasting salad vegetables such as lettuce and cucumber. They can also be boiled or steamed, stir-fried, or added to pasta dishes. When used medicinally, they are usually picked later in the season when their pungency is more obvious, just as they are going into flower.

EXTERNAL USE

The oils contained in arugula leaves are similar to those found in mustard, which have a stimulating effect when used locally, helping to speed up the body's healing process and reducing inflammation. They also used to be applied as a poultice to bruises.

HOW TO GROW

Arugula is an annual that grows fast in most soils, but it does especially well in cool climates, when planted in rich, moisture-retentive soil in partial shade. If it is planted in full sun it tends to bolt. Sow seed in midspring to early summer, about ½ in (1.5 cm) deep, with 12 in (30 cm) between rows, every 2–3 weeks from midspring to summer. Thin the seedlings to 6 in (15 cm) apart when they are large enough to handle. Leaves can be harvested regularly, after 6–8 weeks, to encourage fresh leaf growth.

Do not discard the early thinnings, since they are delicious when used in salads. Mature plants can either be harvested whole or individual leaves can be clipped off as needed for "cut-and-come-again" growth.

Do not allow the soil to dry out in hot weather.

Arugula
recipes & remedies

Arugula leaves have been known for thousands of years for their cleansing and stimulating effect on the system. They are best eaten raw in salads, especially as an appetizer before other foods are eaten.

Arugula and garlic vinaigrette

An excellent dish to enhance immunity and clear congestion.

4 oz (110 g) arugula leaves, washed
2 cloves garlic, crushed
3 tablespoons olive oil
1 tablespoon white wine vinegar
Salt and pepper to taste

Thoroughly mix the olive oil, vinegar, garlic, and salt and pepper to taste. Drizzle over the arugula leaves and eat as an hors-d'oeuvre.

Fennel *Foeniculum vulgare*

Fennel is an attractive, statuesque perennial with blue-green feathery leaves and large umbels of flowers and then seeds. It is a native of the Mediterranean and is a familiar sight in wild open spaces and by roadsides all over Europe. The whole plant has a lovely, sweet licorice smell and taste, which has long been valued in the kitchen, and for making liqueurs, perfumes, and medicines. The soft green leaves make a delicious garnish, traditionally used for fish dishes, since fennel is said to counteract the oiliness in fish. The seeds are often used in making a range of bread, cakes, cookies, and biscuits.

FENNEL CAN
HELP TREAT

- *Arthritis*
- *Bronchial congestion*
- *Colic*
- *Coughs*
- *Flatulence*
- *Fluid retention*
- *Heartburn*
- *Indigestion*
- *Menopausal problems*
- *Nausea*
- *Period pain*
- *Poor appetite*
- *Poor lactation*

Fennel was valued as both a food and medicine by the ancient Egyptians, Greeks, and Chinese. The Egyptians used fennel to treat eye problems, and it was said that fennel enabled the eyes to see clearly the beauty of nature. To this day decoction of fennel seed is recommended as a remedy to bathe inflamed eyes, and until about fifty years ago or so it was customary to wash the eyes of a newborn baby with fennel water. According to ancient Greek mythology, Prometheus hid the fire of the sun in a hollow fennel stalk and brought it to earth to benefit the human race. Fennel was said by the Greeks to give men strength; it was fed to athletes to improve their performance, and they used it as their symbol of victory.

The Greeks also used fennel as a detoxifying remedy and diuretic, and prescribed it as a slimming aid. Hippocrates recommended it for stimulating milk flow in nursing mothers, as herbalists still do today. In the Middle Ages, the seeds were chewed to relieve the pangs of hunger, especially while fasting during Lent.

The swollen stem-base of the Italian Florence fennel (*F. vulgare* var. *dulce*) has a crisp and crunchy texture, a mild aniseed taste, and it is delicious.

Earth up fennel regularly from the time when the bulb is the size of a golf ball until harvesting.

INTERNAL USE

It is as a digestive remedy that fennel is best known today. It has a warming and relaxing effect throughout the digestive tract, stimulating the appetite and promoting digestion and absorption, particularly of carbohydrates and fats. It helps to relieve tension and spasm in the gut, and fennel seeds are included in gripe water for babies, helping to relieve colic or soothe an uncomfortable or restless baby. Fennel tea can help to settle the stomach and relieve gas, nausea, indigestion, and heartburn.

As a diuretic, fennel enhances the elimination of fluid and toxins from the system via the urinary system and it can be used to help treat fluid retention and as a remedy for arthritis. Its relaxing or antispasmodic effects extend to the uterus, where it can help to reduce period pain, while its hormone-like action is used to help regulate the menstrual cycle. It is also helpful during menopause. For the respiratory system, fennel can aid expectoration.

HOW TO GROW

Herb fennel (*Foeniculum vulgare*) does best in well-drained, medium-rich soil and full sun. It can be propagated by sowing seeds from late winter to late spring in position, and thinning to 1 ft (30 cm) as soon as the seedlings are large enough to handle safely. The leaves can be harvested continuously during the summer months, and the seeds from midautumn.

Florence fennel (*F. vulgare* var. *dulce*) prefers a well-drained and sandy soil with plenty of well-rotted manure or garden compost and full sun. It should be sown about 1/2 in (1.5 cm) deep, 18 in (45 cm) apart, in midspring and thinned to 1 ft (30 cm) apart when the seedlings are large enough to handle. Keep plants well watered in dry weather and earth them up (*see left*) as they grow. Harvest in late summer.

Fennel
recipes & remedies

Fennel is often eaten with beans, and vegetables such as cabbage and broccoli, to aid their digestion and help prevent gas.

Fennel soup

A soup to aid digestion by priming the digestive tract for the food to follow. As a mild diuretic, fennel can also help to relieve fluid retention.

2 bulbs fennel
2 tablespoons olive oil
2 medium-sized potatoes, peeled
    and diced
1 quart (1 liter) water or vegetable
    stock
Salt and freshly ground pepper

Cut the feathery leaves from the fennel and set them aside. Trim the bulb and cut into chunks. Heat the olive oil in a large saucepan over moderate heat. Add the potatoes and fennel, and sauté, stirring, for 5–10 minutes, until slightly softened. Pour in the water or stock and bring to a boil, then reduce the heat, cover, and simmer the soup for 20–25 minutes, until the vegetables are tender—test a large piece with the point of a knife. Remove the pan from the heat and allow to cool a little, then puree the soup in a food processor or blender until smooth. Return to the rinsed-out saucepan and reheat gently. Season to taste with salt and pepper, garnish with the reserved fennel leaves, and serve.

Lettuce *Lactuca sativa*

Garden lettuce is a cultivated variety of the wild lettuce (*L. serriola* or *L. virosa*) that can still be found growing on chalky soil throughout the British Isles, as well as on uncultivated land from Asia to northern Europe. Lettuce derives its Latin name from the milky juice (*lactis* in Latin) that exudes from the stem. This juice used to be collected once it had oxidized and turned brown. It was then used for its opiate content as a substitute for opium or laudanum to kill pain, and as a sedative. The juice of the garden lettuce, which is less powerful than that of the wild lettuce, was often used in the past to soothe restless infants to sleep.

LETTUCES CAN HELP TREAT

- *Anxiety and tension*
- *Congestion, coughs, and colds*
- *Constipation*
- *Gastritis*
- *Insomnia*
- *Irritable bowel syndrome*
- *Nervous indigestion*
- *Peptic ulcer*
- *Poor appetite*

It is probable that the garden lettuce first originated in Turkey or Iran, though suggestions concerning its origins range from Siberia to North Africa. It is thought to have been cultivated first by the ancient Egyptians around 4500 B.C. and was probably introduced as a food to Britain by the Romans, and by the 16th century there were apparently eight varieties. Lettuce seeds accompanied the early settlers to North America.

INTERNAL USE

The wild lettuce was used to stimulate the appetite, to enhance digestion and liver function, and to remedy constipation. John Parkinson, the 17th-century herbalist, said that "lettuce eaten raw or boyled, helpeth to loosen the belly, and the boyled more than the raw," while the 16th-century herbalist John Gerard wrote that the cultivated lettuce "is very proper for hot bilious dispositions." Lettuce is still used for its calming and relaxing properties, to help relieve nervous tension and insomnia. Throughout history it has been valued for its anaphrodisiac properties and was combined with marjoram for this purpose.

Lettuce can still be used therapeutically for its cooling and anti-inflammatory properties when there is heat and inflammation. It may be helpful for ulceration or spasm in the digestive tract, as in gastritis, peptic ulcers, colitis, and irritable bowel syndrome. Its antispasmodic properties can help to relax tension and spasm

throughout the body. Lettuce has a cooling and moistening action in the respiratory system, helping to loosen and soothe dry, harsh, and irritating coughs and easing the production of phlegm; it is particularly helpful in this respect when taken with garlic or thyme.

Lettuces contain many beneficial vitamins and minerals—including antioxidants beta-carotene and vitamin C, folic acid, calcium, potassium and iron. The antioxidants help prevent damage caused by free radicals, and so protect against degenerative diseases, heart disease, cataracts, and cancer. Lettuces are useful for dieters, when eaten without an oily dressing, since they are low in calories yet high in fiber, and so are quite filling. The darker the leaf, the more beta-carotene and vitamin C it contains—dark-leafed varieties also contain bioflavonoids, which work in conjunction with vitamin C and antioxidants to help prevent cancer.

HOW TO GROW

The four main types of lettuce are butterhead, romaine, looseleaf, and crisphead, and they thrive in the same conditions: rich, well-drained soil with plenty of organic matter. They can be used as an intercrop between larger vegetables such as cabbage and brussels sprouts. Seeds for summer varieties should be sown every few weeks from midspring to midsummer to ensure a continuous supply for summer and autumn, and in mild regions seeds can also be sown under cloches from early to midwinter for harvesting in late spring. They should be sown about ¹/₂ in (1.5 cm) deep in well-prepared ground, with 1 ft (30 cm) between the rows. When the seedlings are large enough to handle, they should be thinned to 4–9 in (10–23 cm) between plants, depending on the variety grown. Thinnings can be transplanted fairly easily. Seedlings grown under glass can be thinned again in late winter to 6 in (15 cm) apart. Harvest lettuces as they are required. For all but the "cut-and-come-again" varieties, pull the whole plant out of the ground and cut off the root for composting.

Lettuce
recipes & remedies

Lettuce can be used to make infusions and soups. Soup is a good way of using up leaves of lettuces that have matured and gone to seed.

Lettuce tea

An old remedy for sleeplessness or constipation. Especially good for those who cannot digest raw lettuce.

3–4 lettuce leaves, washed
I cup (300 ml) water

Simmer the leaves in water for 15 minutes. Strain and drink I cupful hot before going to bed.

Lettuce soup

14 oz (400 g) lettuce, chopped
I pint (570 ml) vegetable stock
2 tablespoons (30 ml) olive oil
I onion, finely chopped
I clove garlic, crushed
8 oz (225 g) potatoes, diced
Salt and pepper to taste
5 fl oz (150 ml) natural yogurt

Bring the stock to the boil. Place the lettuce in a heatproof dish, and add the stock. Heat the oil in a pan, and add the onion, garlic, and potatoes. Fry over a gentle heat for 10 minutes. Add the lettuce and stock, and season. Bring to the boil then simmer for 2–3 minutes. Allow to cool, transfer to a blender and blend to smooth consistency. Re-heat and add yogurt before serving.

Tomato *Lycopersicon esculentum*

Tomatoes were brought to Europe from South America in the 16th century by the Spanish conquistadors. As a member of the poisonous nightshade family (Solanaceae), the tomato was considered to be potentially dangerous. The earliest record of the tomato in Europe is by the Italian botanist Mattiolus, in 1544, who described the yellow-fruited variety that the Italians called *pomodoro*. The red tomato gained a reputation as an aphrodisiac, and it was known as "love apple," perhaps because its deep red color was thought to represent love and passion.

TOMATOES CAN HELP TREAT

- *Anemia*
- *Constipation*
- *Fluid retention*
- *Heart and circulatory problems*
- *Insect bites and stings*
- *Vitamin and mineral deficiency*

INTERNAL USE

Tomatoes are highly nutritious, rich in vitamins A, C, and E, as well as folic acid, iron, and phosphorus. They have long been used to aid the digestion and assimilation of starchy and fatty foods, and as a laxative, helping to cleanse the bowel of stagnant wastes and toxins. They have also been used for their diuretic properties, aiding the elimination of fluids and wastes via the urinary system, so they make a good cleansing food. Tomato juice has been used to help boost energy and vitality and to detoxify the system. The vitamins A and C are particularly helpful when it comes to the prevention and treatment of infections, and to speed the healing process.

The natural antioxidants contained in tomatoes, including beta-carotene and vitamins C and E, help protect the body against damage caused by free radicals and thereby help prevent degenerative diseases, heart and circulatory problems, and cancer, and to slow the aging process. Recent research has also associated eating tomatoes with a reduced risk of cancer and heart disease, which is related probably to one of the carotenoids named lycopene, also present in grapefruit and watermelon. They are thought to help protect against cancer of the stomach, lung, and prostate. Other research has indicated that eating tomatoes regularly may help in the prevention of appendicitis. Unusually, the cancer-preventing properties of tomatoes appear to be enhanced by cooking, which releases the fat-soluble lycopene, and cooking them in a little olive oil is the best of all methods. Fresh tomatoes are obviously the most nutritious form to use, but canned tomatoes and tomato paste contain only a little less vitamin C than fresh.

EXTERNAL USE

Acne sufferers may find relief by rubbing slices of tomato on their pimples, and some people rub tomato leaves on their skin to relieve the irritation of insect bites.

Caution: Tomatoes can cause allergic reactions in some susceptible people, including headaches, urticaria, and joint inflammation. They can also cause indigestion and heartburn. People who suffer from arthritis and kidney stones should avoid tomatoes because of their oxalic acid content. The leaves can produce an allergic reaction, usually in the form of a rash, in some individuals.

HOW TO GROW

Tomatoes can be grown in the greenhouse in pots or outside in garden beds in a sunny location, depending on the variety. Buy young plants or sow seeds indoors from late winter to early spring in 5 in (13 cm) diameter pots and thin out to 1 plant per 3 in (7.5 cm) pot when the seedlings have developed leaves and are large enough to handle. When planting, place plants 2 ft (60 cm) apart in full sun. Stake them before planting and continue to tie the plant to the stake as it grows. Keep the plant regularly watered—if it is left to dry out and then flooded, the fruit will split. From mid to late summer, pick fruit as it ripens.

Pinch out side shoots as they grow and pull off the small shoots that grow between the branches.

Tomato
recipes & remedies

Unripe, or green, tomatoes may be used for some recipes, but they vary nutritionally in that red ones contain four times as much vitamin A as green ones.

Spicy tomato juice

A healthy, warming drink on a cold day.

1 lb (450 g) tomatoes, skinned
1/2 lemon, juiced
1 teaspoon (5 ml) white wine vinegar
1/2 teaspoon (2.5 ml) chili or coriander, chopped
Worcestershire sauce to taste
Salt and pepper to taste

Blend the tomatoes until liquid and sieve. Add the lemon juice, vinegar, Worcestershire sauce, and salt and pepper. Dilute with a little mineral water and chill. Sprinkle with the chopped chili or cilantro before serving.

Tomato and yogurt juice

A nutritional start to the day. The yogurt helps to offset the heating qualities of the tomatoes.

3 medium tomatoes, skinned, liquidized, and sieved – enough to make 5 fl oz (150 ml) juice
1/2 cup (125 ml) non-carbonated mineral water
1 tablespoon (15 ml) plain live yogurt
Sprig mint, bruised
Salt and pepper to taste

Combine tomato juice and mineral water. Stir in the yogurt. Add the mint and season with salt and pepper.

Parsnip *Pastinaca sativa* subsp. *sativa*

This sweet and starchy vegetable is a cultivated variety of the wild parsnip (*P. s.* subsp. *sylvestris*), which is still found growing throughout central and southern Europe as well as in the British Isles, on uncultivated land and in grassy fields. It is thought that parsnips were first cultivated as a vegetable by the ancient Greeks and Romans and were introduced to northern Europe by the Romans. Parsnips were grown by monks in medieval monastery gardens and they were important for providing vital sustenance for the days when eating meat was forbidden.

PARSNIPS CAN HELP TREAT

- *Constipation*
- *Low energy*
- *Vitamin and mineral deficiency*

Parsnips were traditionally eaten during Lent with salt fish and became popular in beer, wine, jam, and cake-making because of their high sugar content. The 16th-century herbalist John Gerard mentioned that his friend, a Mr. Plat, had made bread from parsnips, but Gerard said, "which I have made no tryall of, nor mean to do." Parsnip roots, being sweet and starchy, were valued as nutritious foods for warming and fattening but there was a superstition that if they were left long in the ground they could cause insanity and were known as "madnips."

The seeds of the wild parsnip were made into medicines by apothecaries for their aromatic properties. They were used for ague (malaria) and intermittent fevers. A decoction of wild parsnip root was a valued remedy for a sluggish liver and jaundice. In Gerard's day, parsnips were called "mypes" and were turned into a marmalade, used medicinally to improve appetite and as a restorative for invalids. John Wesley wrote in his *Primitive Physic*: "wild parsnip both leaves and stalks bruised seem to have been a favourite application; and a very popular internal remedy for cancer, asthma, consumption, and similar diseases."

In traditional folk medicine, parsnips have been considered to be energy-boosting due to their sugar and starch content, and parsnip soup can be given to the elderly and to convalescents as a nourishing, digestible restorative. Parsnips are also thought to be good for the stomach and to support the kidneys and are valued for their diuretic and cleansing properties.

INTERNAL USES

Today, the sweet, pungent taste and starchy texture of parsnips makes them a popular vegetable as well as an ingredient of soups and casseroles. Although starchy, parsnips are low in calories and high in fiber and so make a good food for people watching their weight. They are rich in vitamins C and E and potassium, and they also contain some folic acid. Being high in fiber, parsnips help to keep the bowels regular and to prevent diseases of the bowel. As a member of the parsley family (Umbellifereae)—which includes carrots, celery, and parsley—parsnips contain substances called terpenes, which research has shown may help to reduce the spread of cancer cells and to deactivate carcinogenic substances that cause tumors.

HOW TO GROW

Parsnips do best in well-drained, friable soil, which needs to be deep, because they have long roots. Parsnips, which should be planted in mid-spring, are not harvested until early autumn at the earliest, so they stay in the ground for a long time, but do not require much attention. Seeds take up to 4 weeks to germinate so it is a good idea to sow them with radishes sown between the rows and used as markers for hoeing. Lettuces can also be sown between the rows. Seeds should be sown in their final positions in clusters of three at about 9 in (20 cm) intervals, and thinned to one plant per cluster when they are large enough to handle.

Parsnips can either be lifted in autumn and stored in sand, or left in the open and harvested as required during the early winter months.

Parsnip
recipes & remedies

The parsnip's taste improves considerably after being subjected to a few frosts, which convert some of the stored starch into sugar. This makes the vegetable sweeter and more flavorful.

Parsnip purée

Useful for regulating the bowels.

1 lb (450 g) parsnips, washed and
   chopped
4 cups (1.2 liters) water
1 teaspoon (5 ml) olive oil
Salt and pepper to taste

Cook the parsnips in a little water until they are soft. Puree, strain, and add a little olive oil and salt and pepper to taste, with sufficient of the cooking water to make up 4 cups (1.2 liters) again.

Roast parsnips

An easily digested and nourishing food for the elderly and convalescent.

1 medium-sized parsnip per person,
   peeled and chopped into 4 pieces
1 teaspoon (5 ml) olive or corn oil per
   parsnip

Blanch the parsnips in boiling water for 5 minutes. Heat the oven to 350°F (180°C) pour the oil into a baking pan and put the pan in the oven for 5 minutes to heat the oil. Remove the pan from the oven, add the parsnips and cook them for around 40 minutes, turning them several times, until they are brown and crispy.

Beans *Phaseolus vulgaris*

The snap bean, broad bean, and haricot bean are just some of the cultivars of *P. vulgaris,* which originated in Central and South America and can still be found in mountain regions growing wild. The bean was cultivated in ancient times—seeds were found in deposits at Cuitarrero cave in Peru dating back to 6000 B.C., and in cave deposits in the Tehuacan Valley in Central Mexico dating to 4000 B.C. They were taken North America, and seeds found in New Mexico date to around 300 B.C. In the 16th century, the Spanish conquistadors brought beans to Europe.

There are many different varieties of bean, which vary in the way they grow and in the color and texture of their pod and seed. Those with rather papery pods, such as lima beans (*P. lunatus*), tend to be grown simply for their beans; those such as runner beans (*P. coccineus*) with tough pods can be eaten in their entirety when young and as shelled beans when more mature; and those such as snap beans (*P. vulgaris*) with tender, fleshy, and stringless pods are grown mainly to be eaten as whole green beans.

INTERNAL USE

Fresh green beans are an excellent source of nutrition, containing vitamins A, B, and C, protein, folic acid, calcium, copper, phosphorus, iron, magnesium, and zinc. Being high in fiber they promote healthy bowel function. By speeding food through the intestines they help prevent the bowel wall from coming into contact with toxins, irritants, and potential carcinogens for any length of time. Beans also help to stimulate liver and pancreatic function and thus help to control blood sugar levels.

The vitamins and minerals contained in green beans are vital to the normal function of the nervous and immune systems, and so they can help those feeling tired and run down, and help the body combat infection. Beans' diuretic action aids excretion of fluid and toxins, and thereby may help people prone to arthritis and gout as well as to fluid retention.

Dried beans are high in soluble fiber, which helps to regulate the bowels, and can be used to help relieve constipation, hemorrhoids, and other bowel problems related to constipation.

Beans can help reduce low-density lipoprotein cholesterol. Eating a cup of beans a day has been shown to reduce cholesterol levels. Other studies have shown that eating dried beans lowers blood pressure, making them excellent to prevent as well as treat cardiovascular problems.

In addition, beans contain substances known as protease inhibitors, enzymes that reduce the activation of carcinogens in the bowel. Protease inhibitors have been shown to deactivate oncogenes—genetic carriers in normal cells that can lead to cancer. Beans also contain lignans, which are thought to have anticancer properties, and when acted on by bacteria in the bowel they are converted to hormonelike compounds that may reduce the incidence of breast and colon cancer.

HOW TO GROW

Broad beans prefer a slightly sandy soil. For an early crop in temperate regions, sow seeds in autumn 8 in (20 cm) apart in rows 9 in (23 cm) apart. Germination takes 1–2 weeks. The top of the stem should be pinched out as the first pods form. Tall-growing varieties require support: position stakes at 1 ft (30 cm) intervals on each side of the row and string twine between them. Early crops can be harvested from late spring, and later crops, sown in late spring, from early summer onward.

Snap beans like light, sandy soil. Sow them indoors midspring, harden plants off, and plant out when 2–3 in (5–7.5 cm) high, 6 in (15 cm) apart, leaving 1 ft (30 cm) between rows. Or sow outdoors in late spring, 2 in (5 cm) deep and spaced as above. Plants can be harvested from early summer, when the pods are 4 in (10 cm) long. Continuous picking encourages cropping.

Haricot beans are snap beans that have been left to mature and dry on the plant. When the pods have turned a pale brown, pull the plant up, pick the pods and dry them, shelling them when they split.

Bean
recipes & remedies

Dried beans tend to produce gas in those who lack the enzymes to break down complex bean sugars. These sugars are attacked by bacteria in the bowel, producing gas. If dried beans are soaked in water before cooking, they are less likely to have this effect.

Cooked dried beans

Eating a cupful a day, mixed in with other foods if preferred, may reduce harmful cholesterol levels.

1 lb (450 g) pinto beans
2 pints (1.2 liters) water
Salt to taste

Cover the beans with water and soak them for several hours or overnight. Drain and simmer the beans in fresh water until soft. Flavor with salt. Eat with other vegetables, in a casserole, or cold in a salad with vinaigrette dressing.

Pea 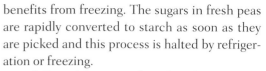 *Pisum sativum*

Peas have been cultivated in southern Europe and in Asia for thousands of years, and eaten either as succulent fresh peas when immature, or in soups and stews when ripe and dried. They are said to be native to the eastern Mediterranean, from Turkey eastwards to Syria, Iraq, and Iran, and were probably first grown in Turkey. The ancient Greeks and Romans were fond of peas, although they did not use them medicinally—apparently they were provided for eating in the pits of theaters (as we might use popcorn). They were introduced to the rest of Europe by the Romans.

PEAS CAN HELP TREAT

- *Constipation*
- *High cholesterol*
- *Low energy*
- *Tiredness and lethargy*

INTERNAL USE

Peas are rich in vitamins A, B complex, and C, as well as the minerals phosphorus, iron, and potassium. They are high in pectin and other kinds of fiber, which not only help to ensure good bowel function but also to control harmful cholesterol levels. It is said there is more fiber in peas than in almost any other food. The edible pea pods are also high in fiber—an excellent food to relieve constipation.

Peas are a member of the legume family (Leguminoseae), and, like their relatives, form a complete protein when combined with grains, whether eaten fresh or dried. Being high in complex carbohydrates, they make an excellent energy-giving food and are particularly good for diabetics, since they help to control blood sugar levels.

Peas also contain substances known as protease inhibitors, which research has indicated have the ability to deactivate certain viruses and carcinogens in the intestines. So peas are potentially good remedies for preventing infections and helping to prevent cancer formation. Studies have linked regular intake of peas with lowered incidence of appendicitis.

Peas were the first foods to be commercially frozen, and they are one of the few foods that actually benefits from freezing. The sugars in fresh peas are rapidly converted to starch as soon as they are picked and this process is halted by refrigeration or freezing.

In folk medicine, peas have long been used as an aid to digestion, to settle the stomach, and reduce heat and inflammation from acidity or gastritis. They have a mild diuretic action, helping to clear toxins from the system.

Caution: Studies indicate the possibility that phytoestrogens contained in peas may reduce fertility. And the purines in peas could precipitate an attack of gout in susceptible people if eaten in large amounts.

EXTERNAL USE

In Germany, children with measles used to be sponged with the water in which peas had been boiled, and a poultice made with cooked peas was applied to boils and abscesses.

HOW TO GROW

Both podded garden peas and snow peas, which are eaten in the pod, like rich, light, well-drained soil. Dig in garden compost or well-rotted manure the winter before planting, and dig the plot again before sowing. Depending on the variety, seeds are sown as an early crop, in midspring, as a main crop in late spring, or as a main crop in summer. Either make a V-shaped groove about 2 in (5 cm) deep, and sow the seeds 2 in (5 cm) apart, or make a flat-bottomed groove about 4 in (10 cm) wide and sow two staggered rows with the seeds 4^{1}/2 in (12 cm) apart in each direction. After sowing, cover the grooves with soil and tread down. Erect support posts and wires or plastic mesh to the correct height for the particular variety grown. Initially, place sticks either side of the groove when the plants are around 3 in (7.5 cm) high, as this will encourage them to climb.

Peas are best picked when young and tender. Feel the pods first to check whether the peas inside have developed. Regular picking will encourage cropping.

Pea
recipes & remedies

Peas and snow peas can be eaten raw and are most nutritious eaten in this way. Alternatively, they can be steamed, boiled or stir-fried, and eaten in soups and risottos.

Pea and pea pod puree

Used to settle the stomach and as an aid to digestion.

1 lb (450 g) peas (including pods)
1 cup (300 ml) water
1 small onion, chopped
Fresh herbs to flavor

Wash the pea pods and place them in a pan with the onion. Add water and simmer with the lid on for 15 minutes, or until tender. Blend in a food processor with some of the cooking water to a rough puree texture. Sieve and add fresh herbs of your choice to flavor.

Pea poultice

An effective remedy when applied to boils and abscesses.

8 oz (225 g) peas, puréed
2 pieces gauze
Light cotton bandage

Place enough pureed peas to cover the affected part of the body between 2 pieces of gauze. Bind it to the affected area with the cotton bandage (see *page 161*).

Radish *Raphanus sativus*

The pungent radish has a long history—it probably originated in China and is recorded as growing there as early as the 7th century B.C. The ancient Egyptians valued it highly for its energy-giving properties, and the slaves who built the giant pyramids were given large amounts of garlic, onions, and radishes to eat. The radish was revered by the ancient Greeks: "such is the frivolity of the Greeks that in the temple of Apollo at Delphi, it is said, the radish is so greatly preferred to all other articles of diet as to be represented there in gold, the beet in silver, and the turnip rape in lead."

RADISHES CAN HELP TREAT

- *Bronchial congestion*
- *Congestion, coughs, and colds*
- *Eczema*
- *Fluid retention*
- *Gout*
- *Liver and gall bladder problems*
- *Poor appetite*
- *Respiratory infections*
- *Rheumatism*
- *Sinusitis*
- *Urticaria*
- *Weak digestion*

INTERNAL USE

There are several different kinds of radish which all have the same medicinal properties, although they vary in degree in relation to their pungency. The long black radish, which has a stronger taste and a more pronounced action than the pink radish, is popular in France and southern Europe, where its ability to help cleanse the liver and invigorate the system is well known. An extract of black radish or freshly expressed juice has often been used to help treat gall bladder and liver disorders such as cholecystitis, gall stones, and hepatic pain. The pink radish was recommended as a folk remedy, to be taken first thing in the morning for liver problems and allergic skin conditions such as urticaria. Radish syrup was prescribed for bronchial congestion, coughs, and whooping cough. Mooli, a long white type of radish, is popular in India and is similarly valued for its reputed liver-regenerative properties and as a remedy for constipation.

As soon as they are chewed, fresh radishes stimulate the flow of digestive juices and saliva in the mouth, whetting the appetite and generally enhancing the digestion. Radishes can therefore benefit those with weak digestion and people who suffer from constipation. Their pungent, stimulating effect is also felt in the respiratory system, where its decongestant action helps to clear bronchial congestion and blocked sinuses and acts as an expectorant. Being rich in vitamin C, they enhance the efforts of the immune system to fight off infection.

Radishes have a mild diuretic action, helping to hasten the excretion of toxins, including uric acid, from the system. This explains their long history of use for aiding the relief of arthritis, rheumatism, and gout, as well as urinary problems, such as gravel, stones and fluid retention.

All types of radish are rich in vitamin C and folic acid, and also contain calcium, copper, iron, potassium, sulfur, and phosphorus. Their pungency helps stimulate the circulation and acts as a tonic to the system by aiding an increase in the efficiency of each cell through the increased blood flow to and from the tissues. As a member of the mustard family (Cruciferae), radishes contain phytochemicals, such as antioxidants and bioflavonoids, which research indicates may be helpful in cancer prevention. The leaves of the radish can be added to salads and can be taken in teas for their diuretic effect.

EXTERNAL USE

Fresh radish pulp can be rubbed on corns and carbuncles to make them disappear.

Caution: Radishes may not suit people who suffer from inflammatory digestive problems, gastritis, or ulcers. They can cause allergies in as they contain salicylate, a compound similar to the active ingredient in aspirin.

HOW TO GROW

Radishes are an easy-to-grow, trouble-free salad crop that can be sown every 2 weeks and does best if grown in light, moisture-retentive soil. Shade them by putting them near taller plants such as parsnips, and keep them well watered in dry weather. Summer radishes should be sown between midspring and early autumn, and winter varieties between mid and late summer, in grooves about $1/2$ in (1.5 cm) deep and 6 in (15 cm) apart. Seedlings should be thinned to 1 in (2.5 cm) apart, and harvested from 4–6 weeks after sowing.

Radish
recipes & remedies

Radishes help to whet the appetite, which is why they are often included in hors-d'oeuvres. They must be chewed well, however, since some people find them hard to digest.

Radish syrup

This can be used to relieve chesty coughs.

3–4 black or pink radishes, washed and thinly sliced
Honey to cover

Sprinkle honey over slices of either black or pink radishes, and strain off the juice after 24 hours. This quantity can be taken at night to relieve symptoms and so encourage a good night's sleep. Large quantities can be stored in the refrigerator in an airtight sterilized jar.

Radish leaf tea

A traditional diuretic remedy for water retention to clear toxins from the system.

2 oz (50 g) radish leaves, washed
2 cups (570 ml) boiling water

Add the radish leaves to the water. Cover the container and let infuse for 10 minutes. Strain and allow to cool. Drink 1 cupful after meals.

Potato *Solanum tuberosum*

The potato was first domesticated and grown by Peruvian Indians in the Andes about 3000 B.C. In the 15th century, the Spanish conquistadors, seeing how highly the potato was valued in South America, took it back to Europe, where other root vegetables, such as carrots and turnips, had long been popular. Legend has it that the potato was first introduced to England in the second half of the 16th century by Sir Francis Drake, who bought a consignment of potatoes in the Colombian port of Cartagena. Others maintain that the potato was sent back to Britain in 1506, from Virginia in eastern North America, by colonists sent out by the Elizabethan explorer, Sir Walter Raleigh.

POTATOES CAN HELP TREAT

- *Acid indigestion*
- *Arteriosclerosis*
- *Arthritis*
- *Chilblains*
- *Colitis*
- *Constipation*
- *Cuts and abrasions*
- *Diverticulitis*
- *Gastritis*
- *High blood pressure*
- *Minor burns and scalds*
- *Stomach ulcers*
- *Sunburn*
- *Ulcers*

INTERNAL USE

Potatoes are high in fiber and carbohydrate and contain proteins, vitamin C (especially when fresh), and minerals—notably potassium—and trace elements. The minerals and vitamins are concentrated in and around the peel, which also has antioxidant properties,that help prevent damage caused to cells by free radicals, thereby helping to protect the body against degenerative diseases, cancer, and the ravages of the aging process. The fiber in potatoes is useful to the health of the bowel, ensuring regularity and helping to prevent bowel diseases, including cancer.

Potato juice has long been popular as a folk remedy. Dr. Vogel, the Swiss naturopathic doctor, maintained that raw potato juice had proved its worth in the treatment of arthritis and was a good remedy for stomach ulcers, relaxing spasm and colic in the stomach and bowel, and helping to reduce excess acid in the stomach. It was traditionally used for indigestion, gastritis, peptic ulcers, liver disorders, gall stones, constipation, and hemorrhoids.

Potatoes may help soothe the urinary system, and they have a mild diuretic action. Being rich in potassium, they help to replace any potassium lost through diuresis. This potassium content is also good for the circulation, and for the regulation of blood pressure, and in fact potatoes are a popular traditional remedy for the heart circulation.

EXTERNAL USE

Raw potato juice has a soothing and an anti-inflammatory action, and encourages healing. It can be applied to relieve skin problems, to treat cuts and abrasions, minor burns, ulcers, chilblains, and sunburn.

Caution: The potato is a member of the nightshade family. The stalks, leaves, green berries, and green tubers of the plant share some of their poisonous properties.

HOW TO GROW

Potatoes are divided into early, midseason, and late varieties. Earlies cannot be stored, but the late types are intended for storing over winter. Potatoes like a sandy soil with a high humus content. Tubers are planted (*see left*) in early spring for first earlies, midspring for midseason types, and about 2 weeks later for late varieties. *Method 1:* Dig a trench 8 in (20 cm) wide and 1 ft (30 cm) deep and stand the sprouted seed potatoes with their tops about 6 in (15 cm) below the ground.

Method 2: Make 4–5 in (10–12 cm) holes, 1 ft (30 cm) apart, and push the potatoes into them, sprout upwards.

Potatoes need to be earthed up, when the shoots are about 6 in (15 cm) high, and this should be repeated every 3 weeks or so. Keep them well watered, watering in the trenches between the rows.

First early potatoes should be ready for harvesting in early to midsummer, midseason types in late summer, and late types in early autumn.

Buy healthy seed tubers for all varieties in mid to late winter and lay in trays, eye-up, in a light, frost-free place to sprout.

Potato
recipes & remedies

Grated raw organic potato and raw potato skins can help to prevent degenerative diseases, including heart disease. When baked or steamed in their skins, potatoes retain much of their nutritional value and are easily digested. Use a juicer to extract the potato juice.

Decoction of potato peelings

This may help to reduce high blood pressure.

Skins of 4–5 organic potatoes
2 cups (570 ml) boiling water

Boil the potato skins in water for 15 minutes, then strain and cool. Drink 1–2 cupfuls of the liquid daily.

Raw potato juice

This is a helpful remedy for both stomach ulcers and arthritis.

Juice of 3–4 organic potatoes
Honey, carrot, or lemon juice to taste

Add the flavoring to the potato juice and drink half a glass (3 fl oz) 4 times a day for 1 month.

Potato lotion

Apply externally for minor cuts, abrasions, minor burns, sores and ulcers, chilblains, and sunburn.

Juice of 3–4 organic potatoes
Olive oil or milk

Mix the potato juice with equal amounts of either olive oil or milk and apply to the affected area.

Spinach *Spinacia oleracea*

The origins of spinach are lost in the mists of time, but it is one of oldest known vegetables and is thought to come from the Middle East, around Persia. The ancient Greeks and Romans grew it, and the Arabs were fond of it and took it to Spain with them in the 10th century, from where it was taken to the rest of Europe. It was grown by the monks in many medieval monasteries in Europe, and was part of many a peasant's diet at that time.

SPINACH CAN HELP TREAT

- Anemia
- Constipation
- Fluid retention
- High cholesterol
- Low immunity
- Poor appetite
- Skin problems
- Tiredness and lethargy
- Vitamin and mineral deficiency
- Weak digestion

INTERNAL USE

Spinach is considered a strengthening and energizing vegetable, perhaps due to its iron content, which enhances oxygenation of the blood; oxygen absorption is futher enhanced by its vitamin C content (although this is hindered to some degree by the oxalic acid contained in spinach).

It makes a good food for those feeling tired, run down, and recovering from illness, and for the anemic and elderly. It is easily digested by most, and has properties that help in enhancing appetite and stimulating the digestion and absorption of food by increasing secretion of digestive enzymes and bile. It is rich in minerals and vitamins, including vitamin C, beta-carotene, iron, folic acid, potassium, magnesium, protein, and chlorophyll.

Spinach has a mild laxative action, helping to clear waste from the bowel, and has diuretic properties, aiding the elimination of fluid as well as toxins. This explains spinach's overall cooling and cleansing action, useful for helping to clear skin problems. The potassium in spinach makes up for any losses caused by increased urination.

Spinach makes a good food to enhance immunity and help fight off infection. The antioxidants beta-carotene and vitamin C contained in spinach help to protect the body from damage caused by free radicals and so help to prevent degenerative diseases, heart disease, and cancer. Spinach may help to lower harmful cholesterol levels. The bioflavonoids abundant in spinach, which impart the dark green color to the leaves, are thought to help deactivate carcinogens and, therefore, inhibit the formation of tumors. Among these bioflavonoids are the carotenoids, beta-carotene and lutein, which have both been shown in recent research to help prevent cancer of the colon, stomach, lungs, and the prostate. Chlorophyll has also been shown to inhibit the action of carcinogens. Folic acid, contained in spinach, not only helps prevent anemia, but is also vital for pregnant women for normal development of the brain and spinal cord of the baby.

Spinach is a good source of protein for vegetarians, but is best when eaten with a grain as it lacks the amino acid, methionine, which prevents it from being a complete first-class protein.

Caution: Due to the high oxalic acid content of spinach, it is contraindicated in gout and arthritis and should be avoided by anyone suffering from kidney or bladder stones.

HOW TO GROW

Spinach needs a moist, fertile soil with a high nitrogen content in order to thrive. It is a cool-season crop, best sown in early spring and late summer. Avoid planting spinach in the hot months, since this is likely to result in poor germination and weak plant growth.

There two types of true spinach —smooth-leaved and the more traditional savoyed, or crinkly-leaf, type. New Zealand spinach (*Tetragonia expansa*) is not, despite its name, a true spinach, although it has a similar taste and does well in warm weather.

Sow seeds 1 in (2.5 cm) deep, leaving 18in (45 cm) between rows. They should be thinned to about 1 ft (30 cm) apart when seedlings are large enough to handle.

When harvesting, do not strip plants entirely. Do not cut the leaves, but pick them by bending them downwards.

Spinach
recipes & remedies

Spinach can be enjoyed as a cooked vegetable or eaten raw in salads. It should be used as soon after picking as possible, since it deteriorates quickly.

Spinach and carrot juice

A nutritious and revitalizing tonic if you are feeling tired or run down.

8 oz (225 g) spinach, washed
3 carrots, washed

Using a juicer, extract the liquid from the spinach and carrots. Drink ¹/₂ glassful (3 fl oz) daily while symptoms persist. Store excess juice in an airtight sterilized container in the refrigerator.

Steamed spinach

Good iron tonic if you are anemic.

8 oz (225 g) spinach, washed
2 teaspoons (10 ml) olive oil
Salt to taste

Steam the spinach lightly for 5–10 minutes. Add the olive oil and salt to taste.

Spanakorizo

Greek-style spinach with rice is delicious, nourishing, and excellent for those with weak digestions, elderly people and convalescents.

3 tablespoons olive oil
1–2 large onions (or 1 bunch scallions), chopped
3 lb (1.3 kg) spinach, stemmed
2¹/₂ cups (600ml) water or vegetable stock
Salt and freshly ground pepper
³/₄ cup (175 g) basmati rice
2 tablespoons chopped fresh dill

Heat the oil in a large saucepan over low heat. Add the onions, cover, and cook for about 10 minutes, or until the onions are translucent but not browned. Wash the spinach and shake off any water left. Add to the onions. Cover and cook for 5 minutes, or until wilted. Pour in the water or stock, add salt and pepper to taste, increase the heat, and bring to a boil. Add the rice, stir well, and cover the pan. Reduce the heat to low and let simmer for 15–20 minutes, until the water is absorbed and the rice is cooked. Stir in the chopped dill. Cover and let stand for 5 minutes before serving.

Whether raw or cooked, fruit is the most delicious form of natural medicine

fruit

available. The taste and smell of organically grown fruit is wonderful, and the simple pleasure of picking an apple or pear from your own tree is hard to beat. Refreshing to the taste buds and cleansing to the system, an excellent source of vitamins, minerals, and natural antibiotics, fruit is the ideal food for eating your way to good health.

Strawberry *Fragaria* x *ananassa*

The sweet and succulent taste of strawberries has been appreciated for thousands of years. The ancient Greeks called the strawberry *komaros*, which means "a mouthful," in reference to its convenient size, while the Romans called it *fragaria* because of its delicate fragrance. Izaak Walton in his *The Compleat Angler* records the quote of Dr. Butler, a 16th-century physician: "Doubtless God could have made a better berry, but doubtless God never did." Its fragrance alone was said to be refreshing to the spirits.

STRAWBERRIES CAN HELP TREAT

- Acne
- Cold sores
- Constipation
- Fevers
- Fluid retention
- Heat rash
- Infections
- Inflammatory eye problems
- Skin problems
- Sunburn
- Vitamin and mineral deficiency

Throughout history the strawberry has been valued not only as a food but also as a medicine. It has long been considered to have cleansing properties, due to a combination of its laxative and diuretic actions, and has been used to purify the blood and as a cooling remedy for hot inflammatory problems. It has also been used for liver problems, inflammatory eye conditions, and to relieve fevers.

INTERNAL USE

Strawberries are a highly nutritious food, rich in vitamins A and C, and in the minerals iron, magnesium, potassium, sulfur, calcium, and silicon, which explain their historical use as a tonic for convalescents. The calcium and magnesium aid normal function of the nervous system, which might help to lift the spirits and calm the nerves. The seeds are rich in pectin and other soluble fibers, providing a mild laxative effect and possibly helping to reduce harmful cholesterol levels. However, the seeds may cause irritation to some people suffering from inflammatory digestive problems such as colitis and diverticulitis.

Recent research has shown that strawberries may help to prevent cancer and degenerative diseases. The antioxidant vitamins A and C help to stop damage caused by free radicals, while the polyphenols contained in strawberries help to combat cancer. The bioflavonoids, including anthocyanin and ellagic acid, have also been shown to help prevent some cancers. Strawberries also have the ability to inhibit the formation of nitrosamines, which are potent carcinogens formed in the intestines under certain chemical conditions.

Strawberries are a helpful food for the immune system and have been shown to have antibacterial and antiviral actions, helping to inhibit the polio virus and Coxsackie virus.

EXTERNAL USE

Rubbing a strawberry on a cold sore may well prove an effective remedy, and strawberries can also be used externally for skin problems such as acne, heat rash, and sunburn.

Caution: Strawberries can cause serious allergies in susceptible people as they contain salicylates, a compound similar to the active ingredient in aspirin. Their oxalic acid may cause irritation to the bladder and kidneys in some people and inhibit absorption of nutrients, such as iron and calcium.

HOW TO GROW

Strawberries like a sunny position, slightly acid, well-drained soil, and plenty of organic matter. They can be grown in beds or containers. Plant

When the berries form, spread straw around the base of each plant to keep the fruit off the ground. Plastic sheets may be used instead of straw.

in late summer, not more than 18 in (45 cm) apart, in rows 3 ft (90 cm) apart. Dig a hole about 2 in (5 cm) deeper than the root system for each plant. Water the plants in and keep them watered in dry weather. Pinch off the runners as they form (do not layer them to raise more plants; it is better to buy new ones). Pick the fruit by the stalks to avoid crushing and bruising when ripe in summer. Remove the straw (*see left*) and cut off the leaves after cropping. Plants will need to be renewed every 5 years.

Strawberry
recipes & remedies

The best way to use strawberry fruits medicinally is simply to eat them, or apply them to the body, fresh. Strawberry leaves, however, like the fruit, have cooling and astringent properties, and they are also used medicinally, especially for diarrhea, kidney and bladder infections, fluid retention, and as a gargle and mouthwash for sore throats and mouth ulcers.

Strawberry leaf tea

This can either be drunk or used as a gargle.

2 cups (570 ml) boiling water
1 oz (25g) rinsed, fresh strawberry
    leaves

Pour the boiling water over the leaves in a teapot and let infuse for 10 minutes. Strain. Drink 1 cupful 2–3 times daily.

Apricot *Prunus armenaica*

This richly colored, velvet-textured fruit is a native of Central Asia and can be found growing wild in China. Apothecaries in the 16th century were familiar with its therapeutic uses: the herbalist John Gerard wrote "being first eaten before the meat they easily descend and cause other meats to pass down the sooner." He was referring to the laxative properties of apricots, which are particularly apparent in apricots from the Hunza area of the Himalayas, where some people apparently live for over 120 years.

APRICOTS CAN HELP TREAT

- *Anemia*
- *Anxiety and tension*
- *Bowel disorders*
- *Constipation*
- *Loss of appetite*
- *Nervous indigestion*
- *Tiredness and lethargy*
- *Vitamin and mineral deficiency*
- *Weak digestion*

INTERNAL USE

Apricots are high in fiber and low in calories, making them an excellent food for those suffering from constipation and those watching their weight. Dried apricots make a good alternative for sweet or fatty snacks, being naturally sweet, and have the added bonus of being highly nutritious. They are rich in vitamins A, B, and C, and contain protein, magnesium, phosphorus, iron, calcium, potassium, sulfur, and manganese; they are particularly rich in iron when dried, and contain plenty of vitamin C to ease its absorption. So apricots make a good, easily digested, and nutrient-rich food for those feeling weak and run down, anemic, or recovering from illness or stress, for pregnant women, children, and the elderly. The vitamins and minerals provide raw materials for the immune system and thereby enhance immunity and the body's fight against disease. The calcium, magnesium, and potassium are all essential for normal function of the nervous system and for the muscular system, and help to support the body through times of stress. They have long been used as a remedy for anxiety, tension, physical and emotional weakness, depression, and insomnia.

Apricots have a beneficial effect in the digestive tract. As well as their laxative effect, they stimulate appetite, promote digestion and absorption, and soothe irritation. Apricot marmalade is an old English remedy for relieving nausea and nervous indigestion. By relieving constipation they help to protect against bowel disease such as diverticulitis and cancer. They also help to protect against cancer in other ways: their antioxidant vitamins A and C prevent damage caused by free radicals, help to protect against heart and arterial disease, and cancer, and to slow the aging process—as demonstrated by the people of Hunza. The beta-carotene in apricots has been shown to offer some protection against cancer of the lung and possibly the pancreas, the skin, and the larynx, or any cancer linked to cigarette smoking.

Apricot kernels contain a substance known as laetrile or amygdalin, sometimes called vitamin B_{17}, which has been said to help the body to fight cancer. It is a controversial remedy as apricot kernels also contain cyanide, which is, of course, poisonous.

Caution: Apricots can cause allergies in some sensitive people as they contain salicylate, a compound similar to the active ingredient in aspirin.

HOW TO GROW

Apricots like a warm, sunny wall or a greenhouse and can be planted either freestanding, if they are in a well-sheltered place, or fan-trained against a sheltered, sun-facing wall (*see pages 139 and 142*). Apricots can grow to 8 ft (2.5 m) with a 12 ft (3.6 m) spread for fan-trained plants and 15 ft (4.5 m) for a bush. Apricots can be planted from autumn to spring. Bushes should be planted at least 15 ft (4.5 m) apart, and staked. The soil should be slightly alkaline, well-drained, and not too rich. Water well after planting, then mulch to keep the soil moist. Apricots need protection from frost when flowering, and should be thinned when the fruits are the size of walnuts (late spring). Harvest in midsummer when fruit is ripe. Prune to open up the center of the plant and remove overcrowded branches in summer. For fan-trained trees, rub out buds in spring (*see page 76*) and pinch out the tips of side-shoots in early summer. Apricots do not store well, and are best eaten fresh, dried, or made into preserves immediately after harvesting.

Apricot
recipes & remedies

Apricots can be eaten fresh or dried. It is worth noting, however, that the sulfur used by some commercial packers—but not usually by home dehydrators— can also cause allergic reactions, so it is always best to buy unsulfured dried apricots, which do not look so invitingly orange.

Apricot iron tonic

An excellent remedy for anemia.

2 cups (570 ml) boiling water
9 oz (250 g) fresh or dried apricots
Honey or sugar to taste

Pour the water over the apricots and let infuse for 10 minutes. Strain. Drink 1 cupful twice a day.

Stewed apricots

A tasty and gentle dish to help regulate the bowels.

1 lb (450 g) fresh apricots
1 cup (300 ml) fresh apple juice
Water sufficient to cover apricots

Cover the apricots with water in a pan and let them soak for several hours. Bring the water to the boil, add the apple juice and simmer gently for about 30 minutes, or until the apricots are tender and the liquid has reduced to a syrupy consistency. Eat 1 small bowlful, hot or cold, in the evening.

Cherries *Prunus avium* & *P. cerasus*

Sour cherries, such as the tart morello cherry, derive from the wild cherry *P. cerasus*, and the sweet varieties from its near relative the mazzard or sweet cherry, *P. avium*. The sweet cherry, with its shiny red fruits, seems to have come originally from Asia Minor. In China, the delicate spring cherry blossom symbolizes youth, hope, fertility, and beauty; in Japan, it is the national flower and represents perfection. In Christian symbolism, cherries were a fruit of Paradise, an emblem of sweetness and goodness. The fruits and the stems have been respected for their medicinal uses for centuries.

CHERRIES CAN HELP TREAT

- *Anemia*
- *Arthritis*
- *Congestion and coughs*
- *Constipation (fruit)*
- *Diarrhea (stems)*
- *Fluid retention*
- *Gout*
- *Tiredness, lethargy*
- *Urinary infections*
- *Vitamin and mineral deficiency*

According to Culpeper, sour cherries "are more pleasing to a hot stomach, procure appetite to meat, to help and cut tough phlegm and gross humours; but when these are dried, they are more binding to the belly than when they are fresh, being cooling in hot diseases and welcome to the stomach, and provokes urine." The stems certainly have an astringent action, and help to protect the lining of the digestive tract from irritation, inflammation, and infection. Their astringency has a binding effect which can be useful for treating diarrhea when taken as cherry stem tea. The fruits are more laxative and make a good remedy for constipation.

INTERNAL USE

Both the fruits and the stems are valued today for their cleansing properties. Their diuretic action helps the elimination of toxins and excess fluid from the body and can thereby help to clear congestion and phlegm, as Culpeper stated, but can also be particularly effective in the treatment of gout and arthritis. They soothe inflammation of the urinary tract and, taken as a cool tea, can help relieve cystitis and other urinary infections.

Cherries, particularly when eaten fresh, are nutritious, rich in vitamins A and C, the anti-oxidant vitamins, and are a good source of beta-carotene, calcium, magnesium, iron, phosphorus, potassium, and zinc. It is not difficult to see why cherries have long had a reputation for rejuvenating the body and mind, for supporting the nervous system, and for reducing stress. Vitamin C, calcium, and magnesium are vital to normal function of the nervous system, the sugars help to provide instant energy, while the wealth of nutrients enhance energy and immunity. Cherries help the body to combat infections and make a good food for those feeling tired and run down or recovering from illness. Ripe black cherries have long been used as a folk remedy for coughs and sore throats, and are particularly useful for treating children as they taste so good.

Tart cherries, such as the morello, *P. cerasus*, have smaller and more bitter fruit and come from Persia and Kurdistan. The bark of the tree was used as a remedy to reduce fevers and for coughs and congestion. The stems are considered more powerfully diuretic than those of the sweet cherries, but both are used for urinary problems, fluid retention, arthritis, and gout.

The bark of the black cherry (*P. serotina*) is frequently used today in modern herbal practice. It has a sedative effect on the nervous system and particularly on the cough reflex, soothing harsh, irritating, or paroxysmal coughs as in whooping cough and croup. Since it contains prussic acid it is best used only on the advice of a qualified medical herbalist.

HOW TO GROW

Cherry trees need rich soil with lots of organic matter. Both sweet and tart cherries can be grown freestanding or trained against a wall (*see page 139*). A freestanding tree can grow to a height of 27 ft (8 m) and both freestanding and fan-trained trees can grow to a span of 15–20 ft (4.5–6 m). They need to be kept well watered, especially in dry weather.

Sweet cherries fruit on old and young wood. Prune to remove dead wood and crossing branches in summer. When pruning fan-trained sweet cherries, rub out new shoots as necessary in spring (*see page 76*) and pinch out the growing tips of other new shoots in summer. Morello cherries are self-pollinating and fruit on new wood. They need to be pruned after fruiting.

Cherry
recipes & remedies

Sweet cherries are best eaten fresh on their own or added to fresh fruit salads. Morello and other tart cherries are suitable for cooking, especially jams and jellies.

Cherry cough syrup for children

Children will happily drink this delicious tasting syrup.

1 lb (450 g) fresh cherries
Juice of $^1/2$ lemon
1$^1/4$ cups (450 g) honey

Simmer the cherries in a little water until thoroughly soft; strain through cheesecloth, squeezing to extract all the juice. Add the lemon juice and honey and stir until the consistency of cream. Pour into bottles, seal, and refrigerate. Take 1 teaspoon 3 times daily, or as required.

Cherry stem tea

Useful for treating diarrhea and for gout, arthritis, and urinary problems.

1–2 oz (25–30 g) cherry stems
3$^1/2$ cups (1 liter) cold water
8 oz (225 g) fresh cherries

Soak the stems in the water for at least 12 hours. Boil them for a few minutes. Pour onto the fresh cherries. Soak for another hour and strain through cheesecloth. Take 3–4 cupfuls daily.

Plum *Prunus domestica*

Plum trees in the orchard, laden with fruit, are one of the delights of late summer and early autumn days. There are many different varieties of plum—more than of any other genus of stone fruit—but they can be used interchangeably for compotes, sauces, puddings, jams, jellies, and fruit salads. The plum is a close relative of the wild sloe (*P. spinosa*) and of the bullace, both of which are sour and astringent, the damson (*P. insititia*) which comes from Damascus, and the greengage (*P. domestica italica*).

PLUMS CAN HELP TREAT

- *Anemia*
- *Arthritis*
- *Bowel disorders*
- *Constipation*
- *Flatulence*
- *Fluid retention*
- *Gout*
- *Skin problems*
- *Tiredness and lethargy*
- *Vitamin and mineral deficiency*

Plums and their dried form, prunes, have been valued at least since Roman times for their energy-giving and laxative properties. In the 17th century, Culpeper recommended prunes "to loosen the belly . . . to procure appetite . . . and cool the stomach"; a decoction of the leaves has long been a folk remedy for constipation, fevers, and fluid retention. Decoctions of prunes were vital ingredients of laxative medicines, and were the favored laxative for children as their action is not preceded by any griping or discomfort. As a household remedy, prunes used to be cooked in water with wine, cinnamon, and lemon peel; however, they are perfectly tasty cooked on their own with just a little water, and even the juice alone will be effective enough for small children. Through their efficient laxative action, plums and prunes help to keep the bowels healthy and to prevent problems such as diverticulitis and bowel cancer from developing.

INTERNAL USE

Plums and prunes are highly nutritious, rich in vitamins B, C, and E, and in minerals iron, calcium, phosphorus, magnesium, sodium, and manganese. Vitamins C and E have an antioxidant action, helping to protect the body against free radicals and thereby helping to prevent degenerative diseases and to slow the aging process. When dried, prunes not only stimulate the muscles of the large intestine to produce an effective laxative action but they also provide a concentrated source of energy—useful for athletes, walkers, cyclists, mountain climbers, or anyone who wants a high-energy but low-weight food to carry with them.

Plums or prunes have a diuretic effect, helping to reduce fluid retention and so aiding the elimination of toxins via the urinary system. Accordingly, they make a good detoxifying remedy and can help to relieve arthritis and gout. They have a gently stimulating effect on the liver, which augments their cleansing action—making plums useful for clearing the skin, and for symptoms associated with a sluggish liver, such as lethargy, heat, irritability, headaches, and poor digestion. Although prunes can initially provoke gas in some people, plums and prunes can relieve gas and bloating, particularly when this is caused by stagnation of food in the intestine due to constipation.

Caution: Plums can cause allergies in some sensitive people since they contain salicylate, a compound similar to the active ingredient in aspirin. Kernels of the pits contain amygdalin which breaks down to hydrogen cyanide in the stomach, so should be avoided.

If your tree is fan-trained, make sure that you "rub out" with your thumb all those buds that are growing toward or away from the wall, when new growth appears in the spring.

HOW TO GROW

Plum trees need rich soil with lots of organic matter. They can be freestanding or grown up against a wall with supports. Freestanding specimens should be planted 12–15 ft (3.6–4.5 m) apart, and wall-trained trees about 15–18 ft (4.5–5.5 m) apart. Some varieties are cross-pollinating, so check when buying young trees. Trees should be pruned in spring and fruits thinned out when branches are laden in summer, and harvested when they are ripe in late summer or early autumn.

Plum
recipes & remedies

Both fresh and dried plums can be taken in many forms. Fresh plums can be eaten raw or made into jams and jellies.

Plum jam

This jam, rich in iron, calcium, magnesium, and manganese, is a good concentrated source of energy, and it also helps to keep the bowels regular. This recipe makes approx. 6 lb (2.7 kg) jam, about 12 jars.

4lb (1.8 kg) purple plums or greengage plums, halved and pitted
2 cups (450 ml) water
8 cups (1.8 kg) granulated sugar

Place the plums in a preserving pan or large saucepan and pour in the water.

Bring to a boil, reduce the heat, cover, and simmer until the fruit has softened to a purée. Remove from the heat and stir in the sugar until it has completely dissolved. Return to high heat and boil for 20 minutes or so, until the setting point is reached. (To test, drop a teaspoonful of the jam onto a cold saucer and refrigerate for 2 minutes, then see if it wrinkles when pushed with a spoon.) Spoon off any froth that has formed. Spoon into hot, sterilized jars and seal with airtight lids. Process in a boiling-water canner for 15 minutes.

Peach *Prunus persica*

The peach, which is a relative of both the plum and the apricot, has been grown for centuries not only for its delicious and refreshing fruit but also for its beautiful spring blossoms and its medicinal properties, particularly those contained in the leaves and flowers. Its name in ancient Roman times, *persica*, indicates that when the peach first arrived in Europe it was believed to come from Persia; in fact, it originates from China, where the first records of its use date back to 551 B.C. It is thought to have come to Europe before the 1st century A.D. and, by the 16th century, the Spanish had taken peaches to South America.

PEACHES CAN HELP TREAT

- *Anxiety and tension*
- *Bowel problems*
- *Coughs*
- *Cystitis*
- *Fluid retention*
- *Gastritis*
- *Heartburn*
- *Indigestion*
- *Skin problems*
- *Stress-related digestive problems*
- *Tiredness and lethargy*
- *Urethritis*
- *Vitamin and mineral deficiency*

INTERNAL USE

Fresh, raw, unpeeled, washed peaches are a good source of the antioxidant vitamins A and C, helping to prevent damage to the tissues caused by free radicals and the effects of the aging process. Peaches, particularly when dried, provide a rich supply of iron and potassium as well as some calcium, magnesium, and phosphorous. Peaches are easily digested. Being rich in easily assimilable sugars, they provide energy, and being high in fiber and with a moistening quality, they make a good laxative. They are therefore an excellent food for sufferers of constipation, and, because they have the ability to regulate stomach acidity, for those prone to excess stomach acid, heartburn, indigestion, and gastritis. They are cooling and soothing to an irritated gut lining, and with their calming effect on the nervous system, can be helpful for all stress-related digestive disorders—indigestion, gas, gastritis, colitis, irritable bowel, and spastic colon.

Peaches also aid kidney function and have a gentle diuretic action, aiding the elimination of toxins and relieving fluid retention. The juice of peaches will help soothe an irritated urinary tract, and help relieve cystitis and urethritis. Their cleansing action on the system can clear skin problems, improve energy, and give a sense of well-being. They help to prevent atherosclerosis and protect against heart and arterial disease by lowering cholesterol. One of the trace elements that peaches contain is boron, which is said to affect the electrical energy in the brain. Having sufficient boron in the diet helps to maintain mental alertness and enhance mental energy and concentration.

An infusion of dried peach flowers was used until recently, particularly in Europe, as a specific calming remedy for anxiety and nervousness and to calm children's tantrums. The infusion was also recommended to relieve nervous coughs and asthma in nervous children.

EXTERNAL USE

Crushed fresh peaches are said to be an effective beauty aid when applied to the face for a few minutes daily, keeping the skin cool, fresh, and youthful.

Caution: Commercially dried peaches may contain a preservative containing sulfites which can cause allergic reactions. Peaches, like other members of the genus, may cause reactions in those susceptible to salicylate, a compound similar to the active ingredient in aspirin.

Established peach trees should be pruned twice a year; in spring as well as in summer after the fruit has been harvested.

HOW TO GROW

Peach trees can be grown against a wall in temperate regions or as bushes in containers in a greenhouse. They like well-drained, loamy soil. Buy 3-year-old trees and plant in autumn or early winter. Fan-trained trees should be spaced 12–15 ft (3.6–4.5 m) apart, and bushes 15 ft (4.5m) apart. Mulch outdoor plants every winter. Fruit should be thinned in midsummer and harvested late summer.

Peach
recipes & remedies

Peaches, like other stone fruits, can be eaten raw, canned, made into chutneys, compotes, or jams, or juiced.

Syrup of peach flowers

An old recipe recommended for constipation and suitable for adults and children alike.

6 cups (1.7 liters) boiling water
1 lb (450 g) fresh peach flowers
1 1/4 cups (450 g) sugar

Pour the water over the flowers in a pan. Cover and let soak for 12 hours or more. Bring to a boil again, cover, and simmer for 5–10 minutes; strain and add the sugar. Heat gently until the mixture reduces to the consistency of a syrup. Store in a tightly sealed, sterilized jar.

Peach compote

A useful remedy for soothing the stomach.

2 lb (1 kg) ripe fresh peaches
1 cup (250 g) sugar
1 1/4 cups (150 ml) water

Peel the peaches, remove the pits, and cut the fruit in half. Heat the water and add the sugar, stirring until it is completely dissolved. Add the peaches and cook over a low heat for about 15 minutes, or until the peaches are tender.

Pear *Pyrus communis*

The pear tree has one of the loveliest blossoms in spring, and its enticing fruit is sweet and refreshing; the juice is like pure nectar. There are over 3,000 different varieties of pear today, some of which were probably known to the ancient Greeks and Romans. Pears are mentioned in Homer's *Odyssey*, and Pliny describes 39 varieties; by Elizabethan times, 232 varieties were available. They have long been popular, both raw and cooked, particularly in medieval monasteries, and pears baked in syrup were evidently considered food fit for a king, for they were served at Henry IV's wedding feast.

P erry, which is an alcoholic drink much like cider but made from pears, had become as popular as cider in 17th-century England. The 16th-century herbalist John Gerard considered it a good digestive. He wrote: "it comforteth and warmeth the stomach." Culpeper recommended pears to cool the blood, and, when sour, to prevent bleeding and diarrhea through their astringent properties.

PEARS CAN
HELP TREAT

- *Acid indigestion*
- *Arthritis*
- *Colitis*
- *Constipation*
- *Coughs*
- *Cystitis*
- *Diarrhea*
- *Fluid retention*
- *Gastritis*
- *Gout*
- *Heartburn*
- *Irritable bowel syndrome*
- *Stress-related digestive problems*
- *Tiredness and lethargy*
- *Vitamin and mineral deficiency*

INTERNAL USE

Today, we can still use pears to benefit the digestion. They are a good source of pectin and fiber, are easy to digest, and make a good food for regulating the bowels, helpful for remedying both diarrhea and constipation. Their cooling and astringent properties (they contain tannins) are useful in inflammatory conditions of the digestive tract, such as gastritis and colitis, and their low acidity and calming properties help to relieve nervous and acidic stomach problems. Pears, therefore, make a good food for people suffering from nervous dyspepsia, acid indigestion, heartburn, irritable bowel syndrome, and diverticulitis. They are excellent for those prone to allergies, as they are one of the least allergenic foods (when

unsprayed and not treated with preservatives). For this reason, they make a good first food when weaning young babies.

Unpeeled, pears are a delicious source of vitamins A, B, and C, and of minerals and trace elements including potassium, calcium, magnesium, iron, and manganese. Being high in natural sugars, they make a good source of quick energy, particularly when dried. They contain pectin, a soluble fiber that helps to control cholesterol, so pears can contribute to the maintenance of a healthy heart and arteries.

Pears have long been used as a remedy for gout and arthritis. They have a mild diuretic action, helping to clear toxins from the system and aiding excretion of excess fluid, as well as of uric acid. Taken as juice or poached in water, their cooling and cleansing effect is helpful when treating hot arthritic joints, as well as for cystitis and other urinary tract conditions. In the respiratory system, pear has a soothing action, relieving harsh, irritating coughs, particularly if cooked with a little fennel or aniseed.

HOW TO GROW

Most pear trees are not self-fertile so you will need to buy two varieties for cross-pollination to occur. Or you could buy a grafted tree, which has three varieties of pear growing from one stem. Pears can be grown freestanding or up a wall (*see page 139*). They like a sunny, sheltered position with a rich, well-drained, loamy soil. They can be planted in late autumn to early winter. Bush trees should be planted 12–15 ft (3.6–4.5 m) apart, and half-standards 20–25 ft (6–7.5 m) apart. Mulch them well every spring and keep them watered in dry weather. Summer pruning is best for trained trees, but standards need no pruning except where there is old, dead wood and branches crossing over each other. Pears should be picked when ripe in the autumn.

Thin out pears as the young fruits become large enough to turn downward. Each group of fruits should be thinned to 2 pears only.

Pear
recipes & remedies

*Pears should be harvested
when the fruit is mature but
still hard and left to ripen at
room temperature before
being eaten.*

Pear compote

*Benefits the digestion and regulates
the bowels.*

2 lb (1 kg) fresh pears
1 1/4 cups (150 ml) water
1 cup (250 g) sugar

Peel the pears. Leave whole if small
or cut in half if large and remove the
cores. Heat the water and add the
sugar, stirring until it is dissolved.
Simmer the pears in the resulting
syrup for 15–20 minutes. Let cool
before serving.

Pears with fennel

A useful cough remedy.

1 pear – halved, cored, sliced, and
deseeded
1 teaspoon honey
1/2 teaspoon fennel seeds

Put all the ingredients in a saucepan
and cover with water. Bring to a boil
and simmer for 45 minutes, or until
the fruit is tender.

Apple *Malus pumila*

The medicinal benefits of the apple have been well documented since the days of the ancient Greeks and Romans, who considered it to be a universal cure. For years, apples have been used in folk medicine for the treatment of colds, congestion, flu, fevers, bronchial complaints, and heart problems. Cooked apples were considered to be a sedative and an aid to restful sleep. An apple a day really did help to keep the doctor away.

APPLES CAN HELP TREAT

- *Arthritis*
- *Congestion, coughs, and colds*
- *Constipation*
- *Cuts and wounds*
- *Diarrhea*
- *Fevers*
- *Fluid retention*
- *Gastritis*
- *Headaches*
- *Hyperacidity*
- *Indigestion*
- *Irritable bowel syndrome*
- *Liver and gall bladder complaints*
- *Peptic ulcers*
- *Rheumatism*
- *Skin problems*

INTERNAL USE

The apple contains a wealth of vitamins, minerals, and other vital nutrients that are easily digested—in fact, the apple itself aids digestion because its malic and tartaric acids regulate stomach acidity and aid the digestion of protein and fat. It is for this reason that it is traditionally served with such fatty meats as pork and goose. It is said that eating apples helps to dampen the appetite, which is a great bonus for dieters. The apple also has a gentle cleansing action, promoting liver and bowel function, and through its diuretic action it aids the elimination of toxins from the body. The pectin in apples helps to bulk out the stool, making it an effective laxative. Interestingly, apples have been used as a remedy for diarrhea for the same reason. Pectin has also been shown to protect against the adverse effects of pollution by binding to toxic metals such as mercury and lead, and carrying them out of the body.

Recent research has indicated that the apple has even more beneficial properties: fresh apples and homemade apple juice both have an antiviral action, and those who eat apples regularly have been shown to have fewer colds and upper respiratory infections. The tannic acid contained in apples is particularly effective against the herpes simplex (cold sore) virus. Other constituents in apples, called polyphenols, have anti-cancer properties, and apples can also help to lower cholesterol and blood pressure levels.

EXTERNAL USE

The pulp left over from juicing apples can be applied to soothe skin rashes. Baked apples, with their skins removed and mixed with a few drops of olive oil, can also be used for this purpose, and can be applied to minor cuts and wounds to facilitate healing.

HOW TO GROW

There are some dual-purpose varieties of apple available, although most are either dessert (eating) or culinary (cooking) varieties. Each variety may be bought in a number of growth forms (*see page 139*), from dwarf, which is approximately 6 ft (1.8 m) high at maturity, to standard, which can grow to 20–25 ft (6–7.5 m). There are also several supported types available. Some apples are self-fertile but others require another variety to pollinate, so check when buying. It is also possible to buy family trees, which have three or four varieties grafted on one plant. Apples need deep, well-drained, slightly acid soil with lots of organic matter. Buy bare-rooted plants, and plant in a sunny place where they will be sheltered from strong, cold winds and frost. Bush trees should be planted at least 12 ft (3.6 m) apart, or 6 ft (1.8 m) apart for dwarf pyramids. Cordons should be planted at least 3 ft (1 m) apart, with 6 ft (1.8 m) between rows, and espaliers and fans 15 ft (4.5 m) apart. After planting, check the soil regularly around the roots, and firm them down if they have been lifted by the frost. Mulch when the soil is warm in early spring. Ideally, keep a 4 ft (1.2 m) circle free of grass and weeds around the base of the tree. Thin the crop early to mid summer and harvest, according to variety, early to late autumn (*see left*).

When picking, lift the apple in the palm of your hand and give it a gentle twist. If it is ripe, it will come away with the stalk attached.

Apple
recipes & remedies

Both eating and cooking apples are excellent for making purees and jams, but eating apples are better for making juice and apple tea as they are sweeter. However, eating apples may lose their flavor and texture when they are baked, so it is better to use cooking apples.

Baked apples

Baked apples with cloves and cinnamon are a tasty remedy for relieving colds and clearing congestion.

1 tablespoon butter, softened
4 large cooking apples
2 teaspoons (25 g) raw sugar
1/3 cup (50g) seedless raisins
8 cloves
4 teaspoons ground cinnamon
2 tablespoons honey
2 tablespoons water

Core the apples and fill them with the butter, sugar, raisins, cloves, and cinnamon. Place the apples in a baking dish and and spoon the honey and water over them. Place in the oven and bake at 400°F (200°C) for 30–45 minutes, or until soft.

Apple and mint juice

This makes a refreshing drink that is especially good for treating colds.

Simply juice 5 to 6 apples and float fresh mint leaves in the glass.

Sage and apple tea

Useful for digestive problems, especially if stress-related.

5–6 fresh sage leaves
2 cups (570 ml) boiling water
2–3 teaspoons fresh apple juice

Make a cup of strong sage tea by infusing the leaves in the boiling water and leaving for 10 minutes. Add the fresh apple juice, strain, and serve.

Rhubarb *Rheum* x *hybridum* & *R. palmatum*

Rhubarb is a magnificent architectural plant, related to the dock, and a native of China and Tibet, where its root has been valued as a medicine since around 3000 B.C. The dried root of the medicinal *R. palmatum* was first brought to Europe by Marco Polo. *R.* x *hybridum*, the edible garden rhubarb, was brought to Britain from the Volga region of Russia in 1573, also for the therapeutic properties of its root, but was not enjoyed as a fruit or preserve until the 1800s. Its name may be derived from the Greek *rheo*, to flow, because of the plant's purgative properties.

RHUBARB CAN HELP TREAT

- *Constipation*
- *Diarrhea*
- *Infections of the digestive tract*

The medicinal rhubarb *R. palmatum* is used today by medical herbalists and has been used in China for at least 5,000 years as an excellent laxative. It has a bitter taste and a cold quality and is prescribed in Chinese medicine for constipation due to heat, to clear stagnation of undigested food, and to clear excess heat from the body, such as fevers. It is also commonly used in prescriptions for headaches, appendicitis, infectious hepatitis, conjuctivitis, gingivitis, nosebleeds, edema, bacterial infections, and a variety of skin problems. It is a powerful medicine, which should only be used for short periods and on the advice of a qualified herbal practitioner.

INTERNAL USE

The roots of both types of rhubarb contain tannins with astringent properties, so they can be used in small doses to treat diarrhea. They also contain purgative substances called anthraquinones, and larger doses are excellent for constipation. Lord Nelson is said to have taken powdered rhubarb root on his voyages so that he would have medicine for every eventuality: for diarrhea, constipation, irritation of the colon, and for infections. An interesting excerpt from an article in *The Lancet*, dated 3 September

1925, written by a Dr. R.W. Duckett, reads: "acute bacilliary dysentery has been treated in that colony [Nairobi, East Africa] almost exclusively with powdered rhubarb for the past 3 years . . . I know of no remedy in medicine which has such a magical effect."

In small doses, rhubarb root acts as a tonic to the digestion, stimulating appetite and promoting digestion and liver function. The root of the garden rhubarb is similar in action in its fourth or fifth year to that of the medicinal rhubarb, though it is milder.

The stalks of garden rhubarb may have a mild laxative action and have been popular in Europe for improving the appetite and stimulating the liver. They contain plentiful amounts of vitamin C and potassium, as well as some calcium, but this is not absorbed well due to the oxalic acid contained in rhubarb, which blocks calcium absorption. The presence of oxalic acid has meant that rhubarb has long been contraindicated to people who suffer from arthritis and gout and those prone to kidney or bladder stones. The tartness of rhubarb means that most people are tempted to add a lot of sugar or honey to make it palatable, so it can end up being a rather high-calorie food. Adding sweet cecily leaves to the cooking pot will reduce the amount of sugar needed.

For an early crop, rhubarb can be forced in its second year by placing a pot or bucket over the crown in late winter.

Caution: The leaves of rhubarb are poisonous to eat.

HOW TO GROW

Rhubarb plants have a life of about 15 years. They like rich soil with plenty of nitrogen. Before planting, prepare the site with organic matter. Plant rhubarb as crowns in early spring, 2 ft (60 cm) apart. Do not harvest in the first year. From the second year onward, harvest from midspring by pulling, rather than cutting, the stems. After harvesting, cover the crowns with straw or dead leaves.

Rhubarb
recipes & remedies

Rhubarb can be very acidic, so it is a good idea to combine it with warming spices such as cinnamon or ginger.

Baked spiced rhubarb

Useful for keeping the bowels regular.

2 lb (I kg) rhubarb
²/₃–³/₄ cup(175–225 g) sugar
2 teaspoons ground cinnamon
4 cloves

Preheat the oven to 400°F (200°C). Cut off all the leaves from the rhubarb and trim the root ends. Wash the rhubarb stalks and chop them into I in (2.5 cm) pieces. Place in a baking dish and sprinkle with the sugar and spices. Cover and bake until the stalks are tender but not mushy. Let cool, still covered, before serving.

Rhubarb and ginger decoction

An excellent remedy for constipation.

¹/₂ oz (13 g) rhubarb
¹/₂ oz (13 g) fresh ginger root
2 cups (570 ml) water

Place all the ingredients in a pan, cover, and simmer gently for 15–20 minutes. Strain and drink I cupful 1–3 times daily, depending on need.

Black currant *Ribes nigrum*
red currant *R. rubrum*

The original currants were small grapes that came from Corinth in Greece. They were known as corinthians or corinth raisins, then corantes, and eventually currants. The fruit now known as dried currants is still a variety of grape. The name currant was transferred to these members of the Grossulariaceae family—familiar to us as black and red currants—due to the similarity in appearance between them and the little grapes from Corinth. There are more than 140 varieties of currants. Very hardy plants, and easy to grow, they do best in cool, moist, northern climates.

CURRANTS CAN HELP TREAT

- *Arthritis*
- *Atherosclerosis*
- *Congestion, colds, coughs, and chest infections*
- *Constipation*
- *Diarrhea*
- *Fevers*
- *Flu*
- *Fluid retention*
- *Gout*
- *Infections*
- *Insect bites and stings*
- *Measles*
- *Mouth ulcers*
- *Poor appetite*
- *Sinusitis*
- *Skin problems*
- *Sore throats*

INTERNAL USE

Red currants have antiseptic properties and, being rich in vitamin C, help to fight off infection. They can be made into a cooling drink to bring down fevers and to speed recovery from colds, flu, and chest infections. Their digestive properties help to stimulate appetite. Red currants have diuretic properties, aiding the elimination of excess fluid and toxins via the urinary system. They make a good cleansing remedy for inflammation and infection of the urinary tract, as well as for arthritis, gout, and skin problems.

Black currants have similar properties but have more nutritional value with their abundance of vitamin C: 1 cup of black currants has about 260 mg of vitamin C, while the same amount of red currants contains only about 55 mg. Black currants have long been used as a folk remedy for colds and flu, coughs and chest infections, as the vitamin C aids the body's fight against infection. It also helps expel toxins and phlegm from the bronchial tubes.

In addition, the tannins in black currants have astringent properties, helping to dry up secretions throughout the body. They make an excellent decongestant. They have a cooling action, reducing inflammation in the body, helping to bring down fevers. Black currant berries and leaves have a diuretic effect, enhancing elimination of fluid and toxins and were once popular as a remedy for kidney stones. The resulting cooling and cleansing action can be used to good effect for easing arthritis and gout, as well as for clearing skin problems.

Due to their astringent nature, black currants have been used since medieval times as a remedy for diarrhea. They contain bioflavonoids, notably some called anthocyanosides, which have an antibacterial action, active against the types of bacteria that often cause diarrhea and gastrointestinal infection. The crushed skins of dried black currants are powdered and marketed in Sweden as an antidiarrheal drug named *Pecarin*. Also, some research has shown that anthocyanosides protect the arteries against atherosclerosis, and thus help to protect against heart attacks and strokes.

EXTERNAL USE

A fresh black currant leaf rubbed onto insect bites and stings helps to provide swift relief. Black currant rob, made from simmering the berries in water with sugar, is an old country remedy for combating sore throats and fevers.

HOW TO GROW

Currants like a heavy, slightly acid soil and a sunny, sheltered position, although they will tolerate shade. They can be planted in early spring or autumn, pruned in late winter or early spring from their second year onward, and harvested from mid to late summer. The fruit grows on old wood and can be harvested more easily by cutting the whole branch and picking from it.

Please note: Planting currants is prohibited in many areas because they, like gooseberries, carry blister rust, a disease fatal to white pines. Check with your local forestry or agricultural authority before ordering or planting. Additionally, currants do not grow well in warm climates.

Currant
recipes & remedies

Hot black currant juice sweetened with a little honey is a delicious drink for adults and children alike, and a good remedy for fevers and respiratory infections. A tablespoon of black currant jam or jelly in a cup of hot water, taken several times a day, used to be popular as a soothing antiseptic remedy for sore throats and colds.

Black currant vinegar

A good remedy for feverish colds.

1 lb (450g) fresh or frozen black
　　currants
Cider vinegar (see method below)
Sugar (see method below)

Place the rinsed black currants in a bowl and cover with cider vinegar. Let stand for 3 days, then press through a sieve. Measure the juice, and for every pint (600 ml) of fluid add 1 cup (225 g) sugar. Heat gently in a saucepan, bring to a boil, and simmer for 5 minutes, removing any foam that forms. Let cool and then pour into bottles. Add 1 teaspoon to a cup of water and drink as needed to help relieve colds and fevers. Will keep for 6 months. No need to refridgerate.

Black currant rob

You make a rob by reducing the juice of a fruit by boiling it until it has the consistency of a syrup, and then preserving it with sugar. This rob is a good remedy for sore throats.

1/2 cup (60 g) fresh or frozen black
　　currants
2 cups (450 ml) water
Honey to taste

Simmer the currants in the water for 10 minutes, then strain and serve with honey.

Red currant and raspberry ice cream

A delicious and nutritious dessert, rich in folic acid and so particularly recommended for pregnant women.

1 1/4 cups (300 ml) water
1 1/2 cups (350 g) superfine sugar
9 cups (1 kg) stemmed red currants
2 cups (250g) raspberries
juice of 1/2 lemon
3 3/4 cups (850 ml) heavy cream

Pour the water into a saucepan, add the sugar, and stir with a wooden spoon over low heat until the sugar is completely dissolved. Turn up the heat, bring to a boil, and boil for 5 minutes. Remove from the heat and allow to cool. Place the fruit in a large freezerproof container. Stir in the cooled syrup and the lemon juice. Whip the cream until stiff. Fold it into the fruit mixture and transfer to the freezer. Once the mixture starts to harden around the edges, remove from the freezer and whisk well to break up the crystals. Return and freeze until firm. Transfer to the refrigerator about 30 minutes before serving.

Gooseberry *Ribes uva-crispa*

The gooseberry is a relative of the currant and is probably one of the least popular of the dessert fruits; not only does it generally taste very sour (new, sweeter varieties have been developed) but it is extremely unfriendly to harvest because of its abundance of prickles. Some say that its name derives from this fruit's long history of use as an ingredient of a tart sauce to accompany goose—its sourness and acidity compensating for the natural fattiness of goose meat. Others say that the name comes from the old English name for gooseberry, *grozer* or *grosier*.

GOOSEBERRIES CAN HELP TREAT

- *Bowel disorders*
- *Constipation*
- *Low immunity*

The gooseberry has long been reputed to have cooling properties and was prepared in vinegar and used internally and externally to cool hot, inflammatory problems. Gooseberries were given as wine, in a fool, or in extracts to bring down fevers; hence its English country name "fever-berry." Gooseberry jelly was an old remedy for stimulating a sluggish liver and easing the digestion of rich and fatty foods. It was considered a good cleansing remedy because of its effect on the liver and its diuretic action, enhancing elimination of toxins via the bowels and the urinary system. The leaves of gooseberry bushes were used to dissolve bladder stones.

INTERNAL USE

Gooseberries are still eaten today for their beneficial effect on digestion. They are high in fiber, which acts as a laxative and helps to remedy constipation and prevent bowel disease such as diverticulitis and cancer. The soluble fiber (pectin) helps to lower cholesterol levels. If eaten regularly, they help to prevent atherosclerosis, high blood pressure, and heart disease.

Gooseberries are a good source of nutrition, being rich in vitamins A and C, and containing potassium, calcium, phosphorus, and iron. The vitamins A and C have antioxidant properties, helping to protect against damage caused by free radicals and to slow the onset of degenerative diseases including cancer.

Caution: Gooseberries may aggravate arthritic pain in some people.

HOW TO GROW

Gooseberries can be grown as bushes, cordons, or fans and, like black currants and raspberries, they need to be protected from damage by birds and should be grown under netting or in special fruit cages. They thrive best in moist, well-drained soil, and will bear the heaviest crops if they are given plenty of organic matter, but no lime, in a sheltered position, with either full sun or partial shade.

Buy 2- or 3-year-old specimens and plant them autumn to spring. Bushes and fans should be planted 5 ft (1.5 m) apart and cordons 15 in (38 cm) apart. Fruit should be thinned from late spring onward, and harvested in midsummer as it ripens.

Please note: Planting gooseberries may be prohibited in your area because they, like currants, carry blister rust, a disease fatal to white pines. Check with your local forestry or agricultural authority before planting.

In autumn or winter, prune young gooseberry bushes to achieve an open center, since this makes fruit far easier to pick.

Established plants will also need to be pruned in summer. Lateral shoots should be shortened to 5 or 6 leaves (see page 143).

Gooseberry
recipes & remedies

Sweet gooseberries can be eaten raw, but acidic ones are more suitable for cooking and for making into jam. Gooseberry leaf tea (follow the method on p.71) is a diuretic which can help to prevent the formation of kidney stones.

Gooseberry and elderflower jelly

An old remedy for easing the digestion of rich and fatty foods.

6 lb (2.7 kg) gooseberries
20 elderflower flower heads
11 cups (2.7 kg) sugar
3^1/$_2$ cups (1 liter) water

Top and tail and wash the gooseberries. Wash the elderflowers and place with the gooseberries in a large saucepan with the water. Simmer gently for half an hour, or until the fruit is soft. Remove the pan from the heat and stir in the sugar. Bring back to a boil and boil rapidly for about 10 minutes, until it reaches the setting point. Remove any foam and strain through a sieve or cheesecloth. Pour into sterilized jars and seal.

Gooseberry fool

A lovely English summer's pudding that will help to ease digestion of rich and fatty foods; it also has a mild laxative effect. This recipe can also be made with other fruits: rhubarb, strawberries, raspberries, apricots, blackberries.

2^1/$_2$ cups (450 g) topped and tailed fresh gooseberries
superfine sugar, to taste
1^1/$_4$ cups (300ml) heavy cream

Put the gooseberries into a saucepan with a little water and cook over low heat until softened to a puree. Remove from the heat, stir in sugar to taste and allow to cool. Whip the heavy cream until stiff peaks form and fold in the fruit. Chill before serving.

Blackberry *Rubus fruticosus*

All parts of the blackberry have medicinal virtues that have been known since at least the time of the ancient Greeks. Dioscorides recommended the whole plant—bark, leaves, root, and fruit—as a remedy for wounds and ulcers, tender gums, and an acid stomach. The plant is rich in astringent tannins that help to dry up secretions and protect mucous membranes throughout the body from irritation, inflammation, and infection. This explains why the fruit, the bark, and the leaves have long been used to curb diarrhea and dysentery, gastritis and colitis, stomach and bowel infections, and bleeding of all kinds.

BLACKBERRIES CAN HELP TREAT

- Anemia
- Atherosclerosis
- Bleeding gums
- Constipation
- Coughs and colds
- Diarrhea and dysentery
- Flu and fevers
- Gastritis
- Hemorrhoids
- High cholesterol
- Heavy periods
- Minor burns
- Mouth ulcers
- Poor appetite
- Skin problems
- Stomach and bowel infections
- Urinary infections
- Vaginal infections
- Varicose veins

INTERNAL USE

Blackberries are rich in vitamin C which helps the body's fight against infection and stimulates the action of the "mucociliary escalator"—tiny hairs in the bronchial tubes that help to clear foreign bodies and infection from the chest. Combined with the astringent properties of the blackberry, this means that the fruit is excellent for the prevention and treatment of colds and flu, coughs and chest infections, sore throats and tonsilitis, and congestion and sinusitis.

A decoction or tincture of blackberry leaves helps to astringe the lining of the uterus and is used by modern herbalists to reduce congestion in the reproductive tract which contributes to heavy bleeding.

Blackberry preparations of all kinds taste delicious and have the added benefit of enhancing appetite and improving digestion and absorption. Blackberries are a nutritious fruit, rich in vitamins A, B, C, and E, as well as in folic acid, iron, and calcium, most of which (except vitamin C) are not lost during cooking. The pectin (a soluble fiber) helps to ensure regular bowel function and to lower harmful cholesterol levels. Thus, the blackberry can provide a good nutritive tonic for children and the elderly as it is easy to digest and assimilate. The iron and folic acid are excellent for pregnant women, the laxative properties help prevent constipation and the pectin helps to protect the heart and arteries from atherosclerosis.

Blackberries and blackberry leaves have a mild diuretic action, which helps to clear excess fluid and toxins from the system via the urinary tract. They can be used to treat urinary infections and also as a cleansing remedy for skin problems, arthritis, and gout.

Recent research has discovered more therapeutic applications of the blackberry. It has shown that the bioflavonoids which color the fruit help the body to fight infection and, like the vitamins A, C, and E, have a powerful antioxidant action, protecting the body against damage caused by free radicals and helping to prevent degenerative diseases such as cancer.

EXTERNAL USE

A decoction of blackcurrant leaves is useful as a lotion for treating hemorrhoids and varicose veins, skin problems, such as ulcers, abscesses, and boils, for cuts, abrasions, and minor burns, as a gargle for sore throats and, as a mouthwash for ulcers and bleeding gums. As a lotion or douche it can be used for vaginal discharge and infection.

Caution: Blackberries can cause allergic reactions in some people as they contain salicylate, a compound similar to aspirin.

HOW TO GROW

Blackberries like partial shade and slightly acid soil. Plant in early autumn or spring, the earlier the better, 1 ft (30 cm) apart for thorned blackberries, 8 in (20 cm) apart for thornless ones. After planting, cut back all stems to 9 in (23 cm). To increase your existing stock of blackberries, layer the tips of the shoots (*see page 143*). This should be done midsummer to early autumn. Tips will be rooted in 3 to 5 months, when they should be cut from the parent plant.

Pick fruit as it ripens from late summer through autumn. Cut down the canes to soil level as soon as they have fruited. Blackberries fruit on growth that is at least one year old.

Blackberry
recipes & remedies

Blackberries should always be picked from sites well away from traffic since they take up heavy metals and other pollution from car exhaust fumes.

Blackberry cordial

A wonderfully warming drink for bedtime; relaxing and helpful in warding off winter ills. Delicious diluted with hot or cold water, or poured neat over stewed apples, apple pie, ice cream, or yogurt.

2 lb (1 kg) blackberries, or enough to give 2 cups (570 ml) blackberry juice

1³/4 cups (450 g) sugar or 6 tablespoons
    honey
10 cloves
5 slices fresh ginger
1 teaspoon (5 g) ground cinnamon
7 tablespoons (95 ml) brandy

Press the ripe blackberries through a sieve to obtain the juice. Place in a saucepan and add the sugar or honey and the spices. Bring to a boil over a low heat, stirring until all the sugar or honey has dissolved. Simmer for 5 minutes. Let cool. When cold, add the brandy, pour into a sterilized bottle, and seal.

Blackberry leaf decoction

Useful for helping to reduce congestion in reproductive tract, especially heavy periods.

2 oz (50 g) blackberry leaves
2³/4 cups (650 ml) water

Place the leaves and water in a saucepan. Bring to a boil and simmer gently for 10–15 minutes. Strain.

Blackberry syrup

Good for sore throats, irritating coughs, or hoarseness – if you don't have time to make it, try a spoonful of blackberry jam instead.

1³/4 cups (650 g) 1:1 mixture of honey
    and refined sugar
2 cups (570 ml) double-strength infusion
    of blackberry leaves (see *page 160*)

Heat the infusion and the honey/ sugar mixture in a saucepan. Stir the mixture as it starts to thicken and skim off any foam from the surface. Let cool before pouring into a sterilized bottle and sealing.

Raspberry *Rubus idaeus*

The raspberry is said to have originated in southern Europe and East Asia but can be found growing wild in many parts of Europe and North America. It is not only the raspberry's exquisite taste but also its medicinal properties that have been valued for hundreds of years in both Europe and North America. In the past, raspberries were made into wines, cordials, and vinegars for medicinal purposes. Raspberry wine and vinegar were considered good remedies for fevers, and the vinegar was used as a gargle for sore throats. It was often mixed with water and given to children to speed febrile diseases such as chicken pox and measles on their way.

RASPBERRIES CAN HELP TREAT

- *Anemia*
- *Anxiety and tension*
- *Bleeding gums*
- *Congestion, coughs, and colds*
- *Constipation*
- *Diarrhea*
- *Fevers and flu*
- *Fluid retention*
- *High cholesterol*
- *Indigestion*
- *Morning sickness*
- *Mouth ulcers*
- *Nausea*
- *Pain in childbirth*
- *Sore throats*
- *Urinary infections*

INTERNAL USE

Both the fruit and the leaves have astringent properties and can be used to treat diarrhea and congestion. They tone the mucous membranes of both the digestive tract and respiratory system, helping to protect them from irritation and infection. This astringency makes them effective when made into a mouthwash for bleeding gums and mouth ulcers.

For many centuries, midwives encouraged women to drink raspberry leaf tea regularly for the last few weeks of their pregnancy, to ease childbirth. Research that began during World War II has confirmed that the action on the uterus of raspberry leaves and, to some extent, the fruit itself, is not simply an "old wives' tale" but is very efficient at easing contractions and helping to ensure a safe and speedy delivery. This is a remedy that is still highly valued today by women all over the world today, not only to ease the pain of childbirth but also to speed recovery afterward.

Raspberry leaves also tone the mucous membranes throughout the kidneys and urinary tract and are useful for preventing infections and fluid retention during pregnancy. They have digestive properties which can be effective in quelling nausea in pregnancy, and their sedative properties provide support to the nervous system, helping to reduce anxiety.

Raspberries are rich in vitamins A and C, and contain calcium, phosphorus, iron, potassium, and folic acid and the leaves contain potassium, phosphorus, calcium, magnesium, manganese, copper, and zinc. These nutrients are excellent for pregnant women, and because raspberries are easily digestible they are a good food for children, the elderly, and those convalescing.

The fruit is rich in fiber, which helps to prevent constipation, and in pectin, a soluble fiber that helps to reduce cholesterol levels. The vitamins A and C and other substances, notably the bioflavonoids, have antioxidant actions helping to prevent degenerative diseases and cancer. One of these bioflavonoids, ellagic acid, is very useful as it is not destroyed by cooking, so that raspberry preserves and jams are still beneficial.

Caution: Raspberries can cause allergic reactions in some susceptible people as they contain salicylate, a compound similar to the active ingredient in aspirin. The oxalic acid in the fruit can cause kidney and bladder stones in some if taken in large amounts, and it inhibits the absorption of iron and calcium.

HOW TO GROW

Raspberries like well-drained, light-to-medium, relatively acid soil, and do best in a sunny, sheltered position. They need supporting with stakes and wire (*see page 140*). Raspberry roots go down as much as 18 in (45 cm) so the lower soil should be loosened with a fork before planting. Plant the canes in autumn to spring, 15–24 in (38–60 cm) apart, with 4–6 ft (1.2–1.8 m) between rows. After planting, cut the canes down to approximately 9 in (23 cm) to a strong, upward-growing bud. Mulch in early spring, and tie the canes into the wire as they grow. Summer-fruiting raspberries will need protection from birds. Pick the fruit when it is ripe.

Cut the old canes out after fruiting. Thin out the new young canes, leaving 5–6 of the best on each plant, and tie them in. In late winter or early spring cut the tip of each cane off to just above the top wire—about 4 ft (1.2 m).

Raspberry
recipes & remedies

Raspberries are most beneficial eaten raw, although they need to be rinsed carefully first. The leaves can be made into tea (see below).

Raspberry vinegar

A useful remedy for sore throats.

4 cups (1.2 liters) cider vinegar
1 lb (450 g) fresh raspberries
3 1/2 cups (1 kg) sugar

Pour the vinegar over the raspberries and let stand for 24 hours. Strain through a sieve. Put in a saucepan and add 1 3/4 cups (450 g) of sugar for each pint (600 ml) of juice. Heat slowly and stir until the sugar is dissolved. Skim off any foam, pour into a sterilized bottle, and seal. Gargle with half a glassful as necessary.

Raspberry leaf tea

To be taken in the last trimester of pregnancy to prepare for childbirth.

1 oz (25 g) raspberry leaves
2 cups (570 ml) boiling water
Honey to taste
Fresh mint or lemon balm, if required

Rinse the raspberry leaves and place them in a teapot. Pour the boiling water over, cover, and let infuse for 10–15 minutes. Sweeten with honey if required and add a few leaves of mint or lemon balm for flavoring if desired. Drink 2–3 times daily.

Blueberry; bilberry

Vaccinium corymbosum; V. myrtillus

The highbush blueberry (*V. corymbosum*) is famous for the delicious blueberry pie that has permeated American folk culture. Many species grow widely all over North America. Bilberries and blueberries are related to cranberries and grow on bushy deciduous shrubs that bear their purplish-black fruits with their characteristic blue bloom in summer. Both the leaves and the fruits have long been used as medicines, particularly by the Native Americans and the early European settlers. The juice of the berries, which produces an indelible dark blue-purple dye, was used to dye cloth.

BLUEBERRIES AND BILBERRIES HELP TREAT

- *Atherosclerosis*
- *Bleeding gums*
- *Bowel infections*
- *Congestion*
- *Cuts and abrasions*
- *Diarrhea*
- *Fevers*
- *Hemorrhoids*
- *Inflammatory problems*
- *Mouth ulcers*
- *Skin problems*
- *Sore throats*
- *Thrush*
- *Urinary infections*
- *Varicose veins*

INTERNAL USE

Both the leaves and the fruits of the blueberry and bilberry bush are rich in tannins which have an astringent action, helping to dry up secretions throughout the body. They can be used for healing mouth ulcers, bleeding gums, and sore throats. The fresh juice used to be popular as a remedy for typhoid and other bacterial infections of the gut, and in Scandinavia dried blueberries or bilberries have long been used to treat childhood bowel infections. Modern research has confirmed the value of this, finding high concentrations of both antiviral and antibacterial compounds in blueberries. It seems that the tannins are responsible for these actions. Blueberries and bilberries also contain antidiarrheal compounds called anthocyanosides, which are particularly effective at aiding the body's fight against infections.

The astringency of the fruits and leaves, together with their diuretic effects, make them useful for treating urinary problems. Blueberries contain a natural antibiotic substance that prevents bacteria from adhering to, and multiplying on, the walls of the urinary tract, and instead help to flush them out of the system. They also help to make the urine more acidic and inhospitable to infecting bacteria.

Blueberries and bilberries also have cooling properties. Taken as an infusion, or as jelly or jam in hot water, they help to bring down fevers and relieve inflammation. By astringing the mucous membranes throughout the body they protect them from inflammation and infection, so helping to guard against digestive, urinary, and respiratory infections.

Blueberries and bilberries are rich in vitamins A, C, bioflavonoids, and iron. They have long been used to strengthen the eyes and improve night vision, as well as an eyewash for infections and inflammation of the eyes such as conjunctivitis and blepharitis. The vitamins and bioflavonoids have antioxidant properties, helping to protect the body against damage caused by free radicals and delaying the onset of aging.

EXTERNAL USE

A decoction of blueberry leaves can be used as a mouthwash or gargle for infections in the mouth and throat, and in a lotion for infected skin conditions and vaginal infections. Their astringent effect is helpful in lotions for varicose veins and hemorrhoids, for speeding the healing of cuts and abrasions, sores and ulcers, and for skin conditions such as eczema and acne.

Caution: These berries can cause allergies in some people, resulting in swollen lips and eyelids, and in hives.

HOW TO GROW

Blueberries and bilberries like a very acid, well-drained sandy soil and a sunny position that is protected from wind and frost. They are slow to fruit but should be established after 6 years. Mature bushes may reach 6 ft (1.8 m) high with a spread of 4 ft (1.2 m). They should be planted 5 ft (1.5 m) apart, in autumn or spring.

They do not transplant well, so it is best to buy container-grown plants. Do not let them dry out, and water them with rain water. Do not use tap water since this may contain lime. Harvest the berries as soon as they are ripe, from midsummer to early autumn.

Blueberry
recipes & remedies

Blueberries and bilberries can be eaten fresh, but are more usually made into jams, jellies, and compotes.

Compote of blueberries or bilberries

For the treatment of bowel and urinary infections.

2 lb (1 kg) fresh blueberries or bilberries
3 tablespoons honey
1/2 cup plus 2 tablespoons water

Wash the fruit and remove any stems and leaves. Heat the water and honey together in a saucepan, boil rapidly for 3 minutes, and add the fruit. Cook gently over a low heat for 10–15 minutes, or until the fruit is soft. Pour into a dish and serve cold.

Blueberry or bilberry leaf decoction

A good remedy for sore throats and mouth ulcers.

1 oz (25 g) leaves
2 cups (570 ml) water

Place the leaves and water in a pan and bring to a boil. Simmer for 10–15 minutes, strain, and let cool before serving and drinking.

Cranberry *Vaccinium macrocarpon*

The cranberry, a creeping plant with attractive red berries, is a relative of the blueberry and bilberry. It is a native of North America and likes to grow in damp ground and peaty soil near acidic bogs. It can be found growing wild but is now widely cultivated in North America and Europe, the cultivated varieties bearing larger, juicier fruit than the wild. The cranberry, like the gooseberry, is sour and acidic and is not eaten raw, but has traditionally been valued as an ingredient of tart sauces, eaten to complement fatty meats. It often accompanies the Christmas turkey in both Europe and North America alike.

CRANBERRIES CAN HELP TREAT

- *Color and night vision problems*
- *Constipation*
- *Cystitis*
- *Kidney stones*
- *Low immunity*
- *Poor appetite*
- *Urinary infections*

INTERNAL USE

Cranberries have long been valued to improve the appetite and enhance the digestion and absorption of food. The sourness of the berries increases the flow of saliva and other digestive juices and thereby activates digestion. The fiber they contain helps to keep the bowels regular. They have been best known over about the last hundred years, however, as a folk remedy for infections of the bladder, kidneys, and urinary tract, as well as a preventative treatment for kidney stones.

It was thought the acidity of the fruit prevented the formation of stones and inhibited infection. The acidity of cranberries is certainly inhospitable to bacteria in the urine, preventing them from reproducing. Research that has been conducted over the past fifty years has actually shown that bacteria cause damage to the urinary system by their adherence to the walls of the urinary tract. Substances in cranberries prevent bacteria from adhering to the walls in this way, and enable them to be easily flushed away out of the system. It has been shown that the regular consumption of cranberry juice can stop an infection in its tracks, even before any symptoms appear.

A daily dose of 1 or 2 glasses of cranberry juice is enough to prevent both kidney and bladder infections in susceptible people, and 2 glasses daily can help to treat them. It is best, however, to avoid those commercial brands of cranberry juice that have a high added sugar content. Modern research has also indicated that cranberries can certainly help in the prevention of calcium-type kidney stones. In addition to helping prevent bacteria from adhering to the walls of the urinary tract, cranberries may well perform the same action in the mouth and the digestive tract, bringing about equally beneficial results in these parts of the body. Research continues.

Further research into the beneficial properties of the cranberry indicates that it contains plenty of vitamin C and other antioxidant substances that help to prevent damage to the body by the action of free radicals. Thus cranberries help to prevent degenerative disease, such as arthritis, heart disease, and cancer, and they also slow the effects of the aging process. The bioflavonoids contained in cranberries, notably a substance called anthocyanin, also help to protect against tumor formation and to enhance vision, particularly at night.

HOW TO GROW

Cranberry plants require a very acid soil and wet, almost boggy, conditions in order to thrive. Add lots of well-rotted organic matter to the soil and water with rainwater, not tap water. In some regions, tap water may contain lime and this will reduce the acidity of the soil.

Cranberries do not like to be moved, so it is best to buy container-grown specimens and plant them about 2–3 ft (60–90 cm) apart. They are best planted out in garden beds, but they can also be successfully grown in containers in areas where the soil is very limy. If cranberries are planted out, spread sand on the soil around the plants to hold down their creeping branches and to encourage them to root.

Cranberries should be picked in the autumn, before the first frost occurs. They are difficult to pick, so wait until most of the berries have ripened and harvest them all at once. They require little˙ pruning—only tidying up is required.

Cranberry
recipes & remedies

Cranberries are too tart to eat raw, but they blend well with apple juice, which makes a low-calorie, nutritious sweetener.

Cranberry and apple juice

Excellent for the prevention and treatment of urinary infections.

2 apples
2 cups (175 g) cranberries

Cut the apples into quarters, wash the cranberries, and remove their stalks. Extract the juice from the fruit in a juice extractor and serve.

Cranberry sauce

To improve digestion, particularly that of meat and poultry.

1 lb (450 g) cranberries
1 cup (225 ml) water
1/2 cup (125 g) sugar

Wash the cranberries and remove their stalks. Place the cranberries and water in a pan, bring to the boil, and simmer until the fruit is soft. Pass the mixture through a sieve and return it to the pan. Add the sugar and heat it until it is dissolved. Pour into a container, let cool, and seal.

herbs

Both the culinary uses and the medicinal benefits of herbs have been recognized for thousands of years. Versatile and rewarding to grow, herbs can be made into safe, natural preparations. Herbs' marvelous health-giving qualities can help to enhance energy and well-being, whether by a stroll through an aromatic garden, a soak in the bath, or a simple herb tea.

Chives *Allium schoenoprasum*

Chives were popular as a culinary and medicinal herb in China as far back as 3000 B.C., and are thought to have been brought to Europe by Marco Polo. Early European herbalists were suspicious of these pungent-tasting fresh leaves, which they thought might cause "evil vapors" in the head, and chives were unpopular until the Middle Ages, when people started to hang the leaves and flowers in bunches from their rafters to ward off evil spirits.

Chives are the smallest member of the onion family, and like their close relatives, leeks, onions and garlic, they contain sulfur and alliin, and an enzyme allinase, that combine to form allicin. This compound has a valuable antiseptic effect, and it is an efficient antibacterial, antiviral, and antifungal agent. Very often the foods and herbs that were valued in the past for their ability to ward off evil spirits are those with antimicrobial properties that fight off infection. The illnesses we now know to be caused by infections were in the past believed to be the result of invasion by evil forces.

CHIVES CAN
HELP TREAT

- *Anemia*
- *Atherosclerosis*
- *Infections*
- *Low immunity*
- *Poor circulation*
- *Sluggish digestion*
- *Stomach and bowel infections*

INTERNAL USE

If added to soups, salads, and omelettes or eaten with cheese or as a garnish, chives can aid the body's fight against infection. Their pungent taste has a warming and stimulating effect in the digestive tract, enhancing appetite and aiding the digestion and absorption of food. The vitamin C and iron content of chives can be useful in helping to combat infection as well as anemia.

Allicin, which is found in chives and other members of the onion family, has been indicated by modern research to help regulate blood pressure and to be of benefit in reducing low-density lipoprotein (LDL) cholesterol. Although chives contain less allicin than their relatives onions and garlic, they are still able to contribute to a healthy heart and circulation, and to support the body's immune system. They are best used fresh.

Chives look attractive in the vegetable and herb garden, with their purple pom-pom flowers and bright green grass-like leaves. They make excellent edging plants and good companions for vegetables, fruit, and flowers, since their antifungal and insecticidal properties help to ward off pests and disease.

HOW TO GROW

Chives are hardy perennials that can be propagated by sowing seeds in late spring or by planting pot-grown specimens in the spring or autumn. Space the clumps about 1 ft (30 cm) apart. Chives will tolerate most soils but they prefer fertile, moisture-retentive soil and full sun. They need to be kept watered, especially in hot, dry weather, and flower heads should be removed in order to stimulate leaf growth.

Chives can be cut as required from spring to fall. The leaves should be cut to within 1 in (2.5 cm) of the soil.

Divide chives every 2 to 3 years by hand in the autumn or early spring. Replant 1 ft (30 cm) apart in well-manured soil.

Dill *Anethum graveolens*

Dill has attractive, aromatic feathery leaves and umbels of tiny yellow flowers, looking and smelling very much like fennel but smaller in stature. Dill comes originally from the Mediterranean and can often still be found growing wild there. The leaves are delicious chopped into salads, vegetable, meat, and fish dishes alike, enhancing them with their delicate flavor. The seeds are traditionally used as a pickling spice for gherkins and cucumbers.

DILL CAN
HELP TREAT

- *Asthma*
- *Babies' sleeping problems*
- *Colic*
- *Constipation*
- *Coughs*
- *Diarrhea*
- *Flatulence*
- *Indigestion*
- *Muscle tension*
- *Nausea*
- *Poor appetite*
- *Sluggish digestion*
- *Stress-related digestive problems*

The name dill is said to derive from the Saxon word *dilla*, meaning to lull, since dill has an ancient reputation for relaxing infants and young children into a restful sleep. It is a remedy that has particular significance for parents with fractious babies. Its tranquillizing properties have been valued since the days of the ancient Greeks and Romans, and its use in this capacity is even recorded in the Bible. The seeds used to be called "meeting house seeds" since they were chewed during long church services to stop the stomach rumbling. Dill makes an excellent remedy for problems of the digestive system.

INTERNAL USE

The pungent taste and aroma of dill stimulates the flow of digestive juices, whetting the appetite and enhancing the digestion and absorption of food. It also has a mildly warming effect, which acts to release tension and to ensure the proper movement of food and wastes along the digestive tract. Both the leaves and the seeds of dill contain volatile oils that have a relaxant effect on smooth muscles throughout the body. In the digestive tract this also helps to release the tension and spasm contributing to colic and gas, indigestion, and nausea, constipation as well as diarrhea. Dill is a vital ingredient in gripe water, which has been used by generations of mothers to

The tender young leaves of thinned dill seedlings can be used in cooking.

soothe their babies' colic. Traditional gripe water is easy to make: simply add ½ oz (13 g) of bruised dill seeds to 1 cup (225 ml) of boiling water and allow it to infuse. Strain, cool, and offer the baby 1 tbsp (15 ml) whenever it is needed. Gripe water can also be useful for cases of flatulence and indigestion in adults.

Dill has a relaxant effect in the bronchial system that helps sooth harsh, irritating coughs and paroxysmal coughing, as well as asthma.

HOW TO GROW

Dill is an annual, which can be propagated by sowing seeds when frost free at intervals to ensure a plentiful supply of fresh leaves. Thin to 10 in (25 cm) apart when the seedlings are large enough to handle. Dill prefers well-drained soil and full sun and should be watered regularly in dry weather. It needs to be planted well away from fennel in order to prevent cross-pollination. Dill can grow up to about 3 ft (1 m) tall. The leaves are ready for picking approximately 2 months after sowing, but they should be harvested before the plant flowers in July or August (depending on climate). Some flowers should be left to go to seed. Harvest seeds once they turn brown. They can be dried out by spreading them on a tray in a dry, warm place and can be planted the following spring to produce that year's crop.

Chervil *Anthriscus cerefolium*

Chervil is an attractive aromatic plant, native to the Middle East, south-east Europe, and Asia. It is a member of the parsley family (*Umbelliferae*) and has delicate, feathery leaves, umbels of lacy white flowers, and a sweet aniseed flavor and smell. It is often found in traditional cottage gardens, and looks lovely in summer with its pink-red stems. The leaves are delicious when used in salads and sauces.

CHERVIL CAN HELP TREAT

- *Arthritis*
- *Fluid retention*
- *Gout*
- *Hemorrhoids*
- *Poor appetite*
- *Poor circulation*
- *Sluggish digestion*
- *Tiredness and lethargy*

Chervil is popular in French cooking and is one of the ingredients of *fines herbes*. It is delicious added to soups, fish dishes, eggs and cheese, casseroles, and salads, and it can be used interchangeably with parsley. It is best used fresh and added to food or sprinkled on top as a garnish just before serving, since its delicate taste and remedial benefits are easily lost through drying and cooking. It is tasty sprinkled on nettle soup—an excellent "spring cleanser".

INTERNAL USES

Like dandelion, watercress, and nettle, chervil has long been valued as a spring tonic, a cleansing herb to clear the body of toxins after the heavy food of winter. In parts of Europe it is traditionally eaten as a cleansing herb on Holy Thursday as part of Easter preparations, and on the Continent, chervil soup is still eaten in many places on Holy Thursday.

In medieval times chervil was a popular remedy for cleansing the liver and kidneys, enhancing the elimination of water and gravel via the urinary system, stimulating the circulation, and purifying the blood. It used to be given after falls or blows to disperse congealed blood and to help prevent bruising. It is still used today for its mild diuretic action, increasing the elimination of fluid and toxins. It may help people with arthritis and gout by assisting in the excretion of uric acid.

Thin plants to 6 in (15 cm) apart and water regularly in hot weather.

Chervil also has a mildly stimulating effect on the circulation and may help to enhance lymphatic drainage. By increasing blood flow to and from the tissues, chervil improves the body's uptake of nutrition and the removal of toxins, thus producing a sense of wellbeing. This is enhanced by its benefits to the digestive system. Its lovely, sweet, and pungent taste stimulates the appetite and improves digestion and absorption. It is a good herb for people with sluggish digestion, particularly the elderly.

EXTERNAL USE

An infusion of chervil leaves used to be popular as an eyewash for sore and tired eyes, and for skin problems. The washed leaves can be applied as a poultice for bruises and hemorrhoids.

HOW TO GROW

Chervil can be grown as an annual or a biennial. It should be sown at intervals of 3 to 4 weeks from spring to early autumn. Seeds germinate in about 2 weeks and leaves are ready to be cut in 6 to 8 weeks. Seeds sown in autumn in a warm sheltered place will be ready for picking in the following spring. Plant seeds *in situ,* 6–8 in (15–20 cm) apart, in well-drained soil, as chervil seedlings do not transplant well. Chervil does best in partial shade. When cutting, leave some flower heads to encourage self-seeding.

Borage *Borago officinalis*

Borage, a native plant of the Mediterranean region, is sometimes found growing wild in other countries. It has blue, star-shaped flowers, which are much visited by bees, and makes an attractive addition to an herb garden. The young leaves of borage are nutritious and lend a refreshing cucumber taste to salads, while the pretty blue flowers can brighten summer cocktails, fruit drinks, and desserts. Borage flowers also look extremely decorative when candied.

BORAGE CAN HELP TREAT

- *Allergies*
- *Anxiety and tension*
- *Arthritis*
- *Asthma*
- *Childhood infections*
- *Congestion, coughs, and colds*
- *Cystitis*
- *Fevers and flu*
- *Fluid retention*
- *Hormonal problems*
- *Low immunity*
- *Poor lactation*
- *Stress-related conditions*
- *Tiredness and lethargy*

One of the old names for borage was "herb of gladness" as it has an ancient reputation for lifting the spirits and dispelling gloom and despondency. Its name borage, or *borago*, is said to derive from *cor-ago*: *cor* meaning heart, or courage, and *ago* meaning I carry or bring; and from the Latin *burra*, meaning woolly, a reference to the hairy stems and leaves. There is an old saying "a garden without borage is like a heart without courage".

INTERNAL USE

Borage has been used throughout history as a tonic to the heart, to increase strength and vitality, and to clear toxins from the system. Today, borage is still respected for its traditional uses and is added to prescriptions to relieve tension and anxiety, to strengthen the nerves, to lift the spirits, and to restore vitality when feeling run down during convalescence. Borage has special significance for the adrenal glands, the organs of "courage" that secrete adrenaline at times of stress. It helps to support the body and increases its ability to deal with stressful situations.

Borage also has a cooling and cleansing effect on the body. By increasing sweat production and through its diuretic action, it clears heat and toxins from the system via the pores of the skin and the urinary tract. In this way it makes a good detoxifying remedy

Borage leaves can be dried in the oven at lowest setting, but keep a check on them since they burn easily.

for treating arthritis, gout, skin problems, and fevers. It is also useful in the treatment of such childhood infections as measles and chicken pox because it helps to bring out the rash. Its decongestant and expectorant action is helpful when treating colds, congestion, and irritating coughs. Borage has long been used to enhance milk flow in nursing mothers, and modern research into the properties of borage seeds suggests that they are rich in gamma-linoleic acid, a fatty acid vital to the normal function of the hormonal as well as immune system, which may make them useful for hormonal problems, allergies, and arthritis.

HOW TO GROW

Borage is a stout, hairy annual that can be grown by sowing seeds outdoors, 1 ft (30 cm) apart, once the danger of frost has passed. Borage does not like being moved, but some people have had success sowing it indoors in pots earlier in the year and transplanting it. In spring, when seedlings are large enough to handle, they should be thinned to 12–18 in (30–45 cm) apart. Borage prefers well-drained soil, sun or light shade, and grows 2–3 ft (60–90 cm) high. It flowers from midsummer onward and its leaves are ready to be picked about 8 weeks after sowing the seed. It self-seeds freely in the right soil and light conditions.

Calendula

Calendula officinalis

This brightly colored plant with its bold, orange daisy-like flowers is a native of southern Europe and parts of Asia, and it has long been a favorite in cottage gardens. In its native sunny climes, the flower was said to be seen on every calends, the first day of each month of the year—hence its name calendula. The petals are easily plucked for use.

Calendula has been highly valued as a medicine through the ages. The Romans recognized its ability to throw off fevers and infections and medieval monks prescribed it for bowel problems, liver complaints, and insect and snake bites. Doctors in the American Civil War adopted calendula as a styptic, to stop bleeding and speed the healing of wounds, and in the First World War, calendula flowers were used in dressings for battle wounds, both to stop bleeding and as an antiseptic.

CALENDULA CAN HELP TREAT

- Bowel infections
- Candidiasis
- Chilblains
- Colds
- Colitis
- Cuts and wounds
- Fevers and flu
- Fluid retention
- Gastritis
- Hemorrhoids
- Hot flushes
- Low immunity
- Peptic ulcers
- Period pain
- Thrush
- Varicose veins and ulcers
- Warts
- Wounds and skin infections

INTERNAL USE

Calendulas have astringent and antiseptic properties, are rich in carotenoids, and are thought to help the body to fight off a range of infections, including colds, the herpes virus, and pelvic and bowel infections, including enteritis, amoebas, and worms, and such fungal infections as candidiasis. In the digestive tract, calendulas may relieve irritation and inflammation and promote healing of the gastric and bowel mucosa, which is useful for treating gastritis and ulceration of the gut as well as colitis and diverticulitis. The bitters stimulate the appetite, enhance digestion and absorption, and improve liver and gall bladder function.

If calendulas are taken in a hot infusion, they increase circulation and promote sweating, helping to relieve fevers, improve blood and lymphatic circulation, and enabling the body to expel toxins. Their diuretic action is useful here also, aiding elimination of fluid as well as toxins via the urinary system. Calendulas also have significance for the female reproductive system. They may help to regulate menstruation and ease problems associated with the uterus that can cause painful periods, excessive bleeding, endometriosis, and cysts. Their hormonal properties are valuable during the menopause, helping to relieve symptoms such as hot flushes and heavy bleeding.

Caution: Avoid calendula internally during pregnancy.

EXTERNAL USE

Calendulas help to stop bleeding and speed the healing of cuts, and their astringent and antiseptic properties are excellent for helping to heal sores and ulcers, varicose veins and hemorrhoids, minor burns and scalds, chilblains, cold sores, and slow-healing wounds.

HOW TO GROW

Calendula is a hardy annual propagated by sowing seeds in spring 12–18 in (30–45 cm) apart, or by taking tip cuttings in summer and early autumn. It likes well-drained soil and full sun and can grow up to 20 in (50 cm) in height. The flowers should be harvested as soon as they are fully open, and the leaves when they are young.

Deadhead calendula regularly for a continuous supply of flowers throughout the growing season.

Caraway *Carum carvi*

Caraway is an attractive member of the carrot family with bright green, feathery leaves and umbels of small white flowers in summer. It is a native of the northern and central parts of Africa, Asia, and the Middle East and grows in many parts of North America. The seeds that follow the flowers in late summer and early autumn are deliciously aromatic and a popular spice for flavoring breads, cakes, biscuits, and liqueurs. In Germany in particular, caraway seeds have long been used to flavor cheese, cabbage and sauerkraut, soups, rye bread, and other baked goods.

Caraway seeds have been enjoyed as a food flavoring for thousands of years. Seeds have been found in Mesolithic excavations that date back about five thousand years, and we know that they were used by ancient Egyptians, Greeks, and Romans, They were popular in medieval and Tudor England, too, where the roots were cooked and served like parsnips to "warm and stimulate a cold languid stomach", the leaves were frequently added to soups and salads, and the seeds were baked in bread.

INTERNAL USE

Caraway contains an aromatic volatile oil that is responsible for the rather parsleylike smell of the leaves and the spicy pungent taste of the seeds. It has been used traditionally as a remedy for flatulent indigestion and to prevent colic and dyspepsia. It also stimulates the appetite and, by increasing the flow of digestive juices, improves digestion and absorption. A few seeds, fresh or dried, chewed before or after a meal are enough to have a beneficial effect and they make a good remedy for relieving hiccups. Caraway "comfits" were traditionally made in a copper pan by covering the seeds with a syrup. They were also given to children in America to stop them hiccuping in church. Caraway seed tea can help to relieve colic in babies and the tannins in caraway seed have an astringent effect, which can help to curb diarrhea.

Caraway seeds have a warming effect throughout the body as they increase the circulation. They have long been used for symptoms associated with cold, such as tiredness, lethargy, weak digestion, or lowered immunity. They have an invigorating and uplifting effect and their antiseptic volatile oils may help the body's fight against infection. By enhancing digestion they ensure regular bowel function and thereby remove from the gut stagnant food and toxins. Taken in a hot decoction, they can help to relieve colds, coughs, and chest infections. A gargle of the decoction will soothe a sore throat.

CARAWAY CAN HELP TREAT

- *Congestion, coughs, and colds*
- *Colic*
- *Constipation*
- *Diarrhea*
- *Flatulence*
- *Hiccups*
- *Indigestion*
- *Poor appetite*
- *Poor circulation*
- *Sore throats*
- *Tiredness, lethargy*

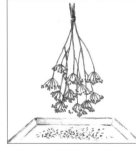

In order to preserve caraway, hang the seed heads upside down over an open container.

HOW TO GROW

Caraway is a hardy biennial which can be propagated by sowing seeds in early to late summer and thinning to 8 in (20 cm) apart when the seedlings are large enough to handle. Caraway likes rich, moderately heavy soil and full sun and can grow to 2 ft (60 cm) in height. The leaves can be harvested at any time, the root in autumn or spring of the second year and the seeds once they are ripe (brown) in late summer or early autumn. Caraway self-seeds freely.

Coriander

Coriandrum sativum

Coriander is a highly aromatic and graceful member of the carrot family, native to southern Europe and the Middle East, and is widely cultivated throughout the world for its parsley-like leaves and spicy seeds. Coriander leaves have a very individual flavor, delightful to many, though some may find it hard to appreciate. The ancient Greeks named coriander after a type of bedbug (*koriannon*) because they considered its taste and smell unpleasant.

CORIANDER CAN HELP TREAT

- *Allergies*
- *Arthritis*
- *Congestion, coughs, and colds*
- *Chest infections*
- *Colic and cramps*
- *Fevers and flu*
- *Flatulence*
- *Gastritis*
- *Indigestion*
- *Inflammatory digestive problems*
- *Peptic ulcers*
- *Poor appetite*
- *Sinusitis*
- *Skin problems*
- *Sore throats*
- *Tiredness and lethargy*

Coriander seeds were some of the earliest to be used in cooking. They are referred to in ancient Sanskrit texts and in the Old Testament as one of the bitter Passover herbs. The ancient Egyptians, Indians, and Arabs used coriander seeds, as they still do today, to flavor curries, meat dishes, vegetables, and soups, as well as in medicines for the digestion. In China, coriander was believed to confer longevity, even immortality, on those who ate it.

INTERNAL USE

Coriander is useful as a mild antispasmodic and digestive. An infusion of crushed coriander seeds helps to relieve gas and indigestion, colic, and cramps, though the seeds are often combined with laxatives to prevent any cramping the latter may cause. Simply chewing the seed stimulates the flow of digestive juices and so enhances appetite and promotes digestion and absorption of food. Coriander is a good remedy for stress-related digestive disorders, such as gastritis and peptic ulcers.

Coriander is an excellent herb to balance the flavor of hot spicy dishes as it has a cooling effect. In Ayurvedic medicine, both the leaf and the seed are added to prescriptions to remedy problems

Coriander should be harvested in early autumn. The seed heads should be cut off and left to dry on trays in a warm place. When they are dry, the seeds can be shaken loose.

associated with excess heat, such as hot inflammatory joint problems, digestive and urinary problems, conjunctivitis, and skin rashes. Fresh coriander leaf tea or juice is used as a remedy for allergies such as hay fever. The fresh leaves are rich in antioxidants, vitamins A and C, as well as in calcium, niacin, and iron, helping to enhance immunity and slow the aging process—just as the Chinese claimed so many centuries ago.

Coriander has an energizing effect on the system when taken regularly, which may account for its recommendation as an aphrodisiac and rejuvenative. The seeds have a reputation for lessening the effects of alcohol by their beneficial action on the liver, and for reducing the soporific effect of large meals. In China the seeds are used to break a fever and to stimulate the appetite.

HOW TO GROW

Coriander is a hardy annual which can be propagated by sowing seeds *in situ* in early spring. Thin the seedlings to 10 in (25 cm) apart when they are large enough to handle. Coriander can grow up to 18 in (45 cm) in height and does best in a sunny position in well-drained soil mixed with plenty of well-rotted manure. Water plants regularly in dry weather to promote the growth of larger lower leaves.

Sweet bay *Laurus nobilis*

Sweet bay, a native of the Mediterranean, is popular in gardens as an evergreen tree or bush, and its shiny aromatic leaves are vital ingredients of a bouquet garni and provide excellent flavoring for soups, stews, casseroles, and marinades. Its mildly pungent taste was once used for flavoring custards and milk puddings. For many centuries sweet bay has been highly valued for its aromatic properties and mythical origins, as well as for its cleansing and antiseptic properties.

BAY CAN
HELP TREAT

- *Arthritis*
- *Bruises and sprains*
- *Congestion, coughs, and colds*
- *Chilblains*
- *Colic*
- *Diarrhea*
- *Fevers and flu*
- *Flatulence*
- *Fluid retention*
- *Gout*
- *Headaches*
- *Indigestion*
- *Muscular aches and pains*
- *Nausea*
- *Period pains*
- *Poor circulation*
- *Rheumatism*
- *Tiredness and lethargy*

In its native Mediterranean, sweet bay can grow as high as 50 ft (15 m), bearing creamy-white blossoms and purple berries. However, in cooler climates it generally reaches no higher than 20 ft (6 m) and rarely flowers.

INTERNAL USE

The medieval herbalist Culpeper recognized the warming nature of bay, which is still valued today to stimulate the circulation and ward off the effects of cold, such as tiredness, lethargy, chilblains, cramps, rheumatism, respiratory ailments, congestion, gas, and stomachaches. Bay's pungent properties are particularly recommended to those who tend to feel the cold and those with weak digestions. Added to food or taken in a hot infusion, it stimulates the appetite and promotes digestion and absorption, easing indigestion, nausea, gas, colic, and diarrhea. The volatile oils that lend bay its characteristic taste and smell have an antispasmodic action, releasing muscle tension throughout the body and helping to relieve cramps, period pains, and headaches. The volatile oils are also antiseptic, helping the body's fight against infection, and are excellent for respiratory infections—colds, sore throats, coughs, and flu. Taken hot, bay reduces

To achieve a ball-headed shape, the tip of the leader (see page 143) should be removed when the tree is 4 ft (1.2 m) high. The lower laterals should be pruned to 3 or 4 leaves. When the tree is mature, headers should be trimmed to 4 or 5 leaves and lower laterals removed.

fevers, and its expectorant properties are helpful for coughs and bronchial congestion; bay has long been used for chronic bronchitis. Bay also has a diuretic effect, assisting the elimination of excess fluid and toxins from the system, which can help to relieve such problems as arthritis.

EXTERNAL USE

Essential oil of bay can be used diluted in a massage oil to improve circulation, which is helpful for chilblains, bruises and sprains, aching muscles, and arthritic joints. A few drops of oil added to a bowl of hot water makes an inhalant for sore throats, colds, and coughs.

HOW TO GROW

Bay can be propagated by taking 4 in (10 cm) cuttings in August or September or by layering established shrubs in late summer or early autumn, but it is easier to buy a plant from a nursery. A bay tree can either be planted in a flower bed and allowed to grow to its natural height or grown in a tub and its height kept down by pruning into a ball-headed shape (*see left*). Bay needs a sunny, sheltered site, protected from wind and frost. In cooler regions it may be summered outdoors in a tub and wintered indoors. The leaves can be harvested all year around.

Lavender

Lavandula officinalis

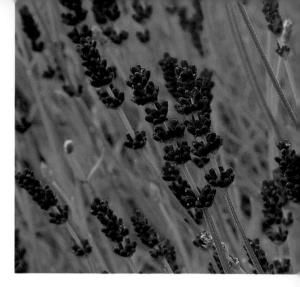

This aromatic shrub, which is a native of the Mediterranean, is one of the most popular scented shrubs in both European and North American gardens. It has been popular since the time of the ancient Greeks, and was used by the Romans to perfume their baths, which explains its Latin name, which comes from *lavare*, meaning to wash.

LAVENDER CAN HELP TREAT

- *Anxiety and tension*
- *Arthritis*
- *Cuts and abrasions*
- *Flatulence*
- *Headaches and migraine*
- *Indigestion*
- *Insect bites and stings*
- *Insomnia*
- *Minor burns and scalds*
- *Muscular aches and pains*
- *Nausea*
- *Nervous palpitations*
- *Poor appetite*
- *Rheumatism*
- *Sinusitis*
- *Sore throats*
- *Stress-related problems*

The medicinal properties of lavender were much appreciated in the 16th and 17th centuries, when one writer observed that lavender was "of especiall use for all the griefes and pains of the head and the heart" and Italian herbalist Mattioli said "it is much used in maladies . . . of the brain due to coldness . . . it comforts the stomach and is a great help in obstructions of the liver and spleen."

INTERNAL USE

Lavender is still considered an excellent remedy for the nervous system and digestion. Its essential oils have a balancing effect on the emotions, calming the mind and lifting the spirits. Taken as a tea or tincture, or by inhaling the volatile oil, lavender makes an effective remedy for anxiety and nervousness, and for stress-related problems such as headaches, migraine, aching muscles, insomnia, tiredness, palpitations, and digestive problems. Lavender also has strengthening properties, restoring energy and vitality to those feeling nervously run down or exhausted.

The plant's relaxing properties can be felt throughout the body. In the digestive tract it releases tension and spasm and can be used to relieve flatulence, nausea, indigestion, and poor appetite. The volatile oils in lavender, as in all aromatic herbs, have an antiseptic effect, helping to combat stomach or bowel infections that cause nausea, vomiting, or diarrhea. Used as an inhalation or in a vapor rub the oil helps to clear the nose and sinuses and to resolve infection in the nose, throat, and chest.

Lavender has a diuretic action and this, combined with its diaphoretic properties (enhances sweating), makes it a good cleanser, eliminating toxins via the pores and the urinary system.

EXTERNAL USE

Lavender can be used to treat cuts, sores and wounds, and to soothe mild burns and scalds. It speeds healing by stimulating tissue repair and helps minimize scar formation. When the oil is applied undiluted it can relieve pain, and it speeds the healing of burns, scalds, and insect bites and stings. The oil can be diluted in massage oils and rubbed into painful joints, aching muscles, and bruises. The tea is a good gargle for sore throats and a mouthwash for ulcers and inflamed gums.

Lavender leaves can be picked at any time, but the stems should be harvested as soon as the flowers open. They can be dried by hanging in small bunches in paper bags.

HOW TO GROW

Propagate by sowing in late summer or autumn, or take 4–8 in (10–20 cm) stem cuttings in spring or summer. Thin or transplant to 18 in–2 ft (45–60 cm) apart, or 1 ft (30 cm) apart to make a lavender hedge. Lavender can grow up to 3 ft (90 cm) high and flowers in midsummer. It likes full sun and well-drained, rather sandy soil and should be pruned in spring or late autumn to prevent straggly growth.

Chamomile

Matricaria recutita & *Chamaemelum nobile*

There are two types of chamomile used medicinally: first, the annual German chamomile (*Matricaria recutita*) and, second, the perennial creeping Roman chamomile (*Chamaemelum nobile*). The medicinal properties of the two plants are almost identical, but German chamomile is generally preferred since it tastes less bitter.

The principal constituent of both the German and Roman types of chamomile is a blue volatile oil containing azulenes, and it is these that lend chamomile its distinctive apple-like fragrance.

CHAMOMILE CAN HELP TREAT

- *Anxiety and tension*
- *Arthritis*
- *Babies' colic*
- *Colitis*
- *Conjunctivitis*
- *Cuts and bruises*
- *Cystitis*
- *Earache*
- *Eczema*
- *Gastritis*
- *Hyperactivity*
- *Insomnia*
- *Migraine*
- *Minor burns and scalds*
- *Nausea*
- *Neuralgia*
- *PMS*
- *Sore throats*
- *Stomach and bowel infections*
- *Stress-related digestive problems*
- *Thrush*
- *Urticaria*

INTERNAL USE

The blue volatile oil contained in chamomile has an anti-inflammatory action, helping to relieve inflammation as in for example gastritis, colitis, and irritable bowel syndrome. However, chamomile is probably better best known as a relaxing herb. It reduces tension and anxiety, relieves insomnia, and is excellent for tense, stressed people, who tend to be over-sensitive, irritable, or hyperactive and who are also prone to digestive problems and allergies. It can be used for soothing over-excited children and for encouraging quiet and sleep. Chamomile relieves digestive discomfort and soothes colic in young babies and has analgesic properties that can help to relieve such discomfort as that caused by teething. Chamomile's analgesic action helps to reduce pain in, for example, migraine, neuralgia, muscle tension, arthritis, and earache.

Chamomile is also used to treat a range of respiratory infections and fevers, in both adults and children. It has an antiseptic action and by helping relaxation and aiding sleep, it encourages the body's natural recovery processes through rest. It can also reduce the symptoms of PMS, and can be given for mastitis and menopausal changes. During pregnancy it can help to quell nausea, and if it is taken throughout childbirth it can relax tension.

EXTERNAL USE

Chamomile can help to reduce inflammatory skin reactions such as eczema and urticaria, minor scalds and burns, cuts, wounds, and ulcers. Chamomile tea makes a good antiseptic lotion for the skin, a mouthwash and gargle for throat infections, a douche for vaginal problems such as thrush, an eyewash for conjunctivitis and inflamed eyes, and chamomile can also be added to bath water to soothe cystitis. The essential oil can be used in massage oils to relieve the pain of arthritis, neuralgia, and muscular aches and pains. It also has the ability to speed tissue repair.

HOW TO GROW

German chamomile is an annual plant that can be propagated by sowing the seeds *in situ* in the spring or autumn in well-drained, preferably chalky soil. Thin seedlings to about 8–10 in (20–25 cm) apart. German chamomile flowers throughout the summer months and self-seeds freely. It grows best in full sun and can reach 2 ft (60 cm) in height.

Roman chamomile is a perennial and you can propagate this creeping plant by sowing seeds *in situ* in spring 12–18 in (30–45 cm) apart, taking 3 in (8 cm) cuttings in summer or dividing plants in the spring or autumn. It grows best if it is planted in light, well-drained soil and in full sun. The leaves can be gathered at any time, and the flowers once they are fully open. Keep it free from weeds and remove dead flowers frequently. For a chamomile lawn, plant seedlings about 4–6 in (10–15 cm) apart.

Lemon balm

Melissa officinalis

Lemon balm has a delightful, refreshing lemony scent and flavor, which is excellent for adding to cooling summer drinks and fruit cups and is delicious as a tea. The fresh chopped leaves can enhance a leafy salad and enliven jellies, jams, and desserts. Lemon balm has been favored not only for its sweet taste but also for its many medicinal virtues since at least ancient Roman times. The plant's botanical name, *Melissa*, is the Greek word for bee, and when lemon balm plants are introduced near hives they will help to attract new members to the colony.

LEMON BALM
CAN HELP
TREAT

- *Allergies*
- *Anxiety and tension*
- *Childhood infections*
- *Cuts and abrasions*
- *Eczema*
- *Hay fever and allergic rhinitis*
- *Indigestion*
- *Inflammatory eye problems*
- *Insect bites and stings*
- *Muscular aches and pains*
- *Nausea*
- *Period pains*
- *PMS*
- *Viruses*

The modern use of lemon balm bears out the wisdom of the ancients. Recent research has shown that it influences the limbic system in the brain, which is concerned with mood and temperament, and explains why it has therapeutic benefits in the treatment of anxiety and depression. While having a relaxing and sedative effect, enhancing relaxation and aiding sleep, lemon balm is also rejuvenating and can be used to aid concentration and memory.

INTERNAL USE

Lemon balm has particular significance for the digestive tract, relaxing tension, and soothing irritation and inflammation. Lemon balm tea will help to relieve indigestion, heartburn, nausea, gas, and diarrhea, and is excellent for stress-related digestive problems. It is an easy herb to give to babies and children since it has a lovely lemon taste, and will not only calm excitable children at night, but it can also relieve colic and stomach upsets and reduce fevers. Its antiseptic properties will help to fight off invading microbes in some childhood infections. Lemon balm has been shown to have an anti-viral action

After the first year of growth, cut the stems back to approximately 6 in (15 cm) from the ground in June. They will need to be cut again in October (see right).

effective against herpes simplex (cold sores), mumps, and other viruses, due to the polyphenols and tannins in the leaf. The volatile oil also has an antihistamine action, which is helpful for treating hay fever. The relaxing effect of lemon balm may help to relieve period pain and PMS, and if it is taken in the weeks before childbirth it helps lessen pain during the birth.

EXTERNAL USE

Lemon balm can be used for skin problems such as eczema and for inflammatory eye problems, and its crushed leaves can be applied to soothe insect bites and stings. A lotion made from the tea speeds healing and helps to stop infection in cuts.

HOW TO GROW

Lemon balm is a hardy perennial. It can be propagated by sowing in spring, or when frosts have passed, in moist soil and full sun and thinning seedlings to 18 in (45 cm). It should be watered in dry weather. Lemon balm can grow to 2–3 ft (60–90 cm) high and flowers in midsummer. It can be harvested throughout summer and plants should be cut back to just above ground level in October.

Mint *Mentha* sp.

Many of the different species of this delightfully refreshing and aromatic herb can be used interchangeably —these include peppermint, lemon mint, spearmint, apple mint, and pineapple mint. Although they all have broadly similar medicinal properties and culinary virtues, the action of peppermint is the most pronounced. The mildly pungent taste and smell of this herb have been enjoyed for thousands of years all over the world, not only in the garden but also in the kitchen and the world of perfume and confectionery.

MINT CAN HELP TREAT

- *Asthma*
- *Back pain, neuralgia*
- *Cold sores*
- *Congestion, colds, coughs, and chest infections*
- *Diarrhea*
- *Fevers*
- *Gingivitis*
- *Headaches and migraine*
- *Joint pain*
- *Mouth ulcers*
- *Nausea*
- *Period pain*
- *Ringworm*
- *Sinusitis*
- *Skin infections*
- *Sore throats and tonsillitis*
- *Tiredness and lethargy*

The volatile oils in mint, notably camphor and menthol, have an effective antiseptic action which was valued in the past for treating infections such as cholera, TB, diphtheria, and dysentery. Today, research has confirmed the antibacterial, antiviral, antifungal, and antiparasitic properties of the volatile oils in mint, which make it a very good remedy for colds, flu, sore throats, tonsillitis, coughs, and chest infections.

INTERNAL USE

Mint is excellent for enhancing the appetite and promoting the digestion and absorption of food. It has valuable antispasmodic properties, relaxing tension in the digestive tract that might interfere with good digestion, and easing gas and distension, colic, indigestion, nausea and heartburn, constipation, and travel sickness. The astringent tannins found in mint help to protect the lining of the gut from irritation, inflammation, and infection, and make it an excellent remedy for diarrhea, bowel infections, and spastic constipation.

Mint's antispasmodic action, coupled with its beneficial relaxing effect on the nervous system, means that it can be used for tension and spasm throughout the body—for asthma, period pain, headaches, muscle tension, and insomnia. Its analgesic properties also help to relieve pain: migraine, back pain, sciatica, arthritis, and gout all respond well when mint is taken internally and applied locally as a lotion. Though relaxing, mint also has a revitalizing effect, dispelling tiredness. This is because of its stimulating effect on circulation, enhancing the nutrition of every cell and ensuring good blood flow to the brain. Taken in a hot tea, mint disperses blood to the surface of the body and causes sweating, making it a first-rate remedy for fevers and for cleansing the body by enhancing elimination of toxins via the pores. Its astringent and decongestant action helps to relieve complaints such as congestion, which is why menthol is often found in pharmacists' decongestant preparations.

EXTERNAL USE

When applied to the skin, mint can help to treat ringworm and herpes simplex (cold sores). It can be used as a gargle for sore throats, and as a mouthwash for ulcers and gum problems. As a lotion, it can be applied to help back pain, sciatica, arthritis, and headaches.

HOW TO GROW

Mint can be propagated by dividing roots and planting them 10 in (25 cm) apart in autumn or spring, or by obtaining a few pots from a nursery and planting them outdoors 6 in (15 cm) apart. Mint does best in sun or light shade, and it will grow under almost any conditions in almost any soil. It grows up to 3 ft (90 cm) high and flowers in midsummer. It can be harvested throughout the growing season. If you are planting mint in a flowerbed, it is best in a pot or bucket (with drainage holes) sunk into the ground to prevent its underground runners from spreading too far.

Basil *Ocimum basilicum*

Basil, with its delicious, spicy, clove-like fragrance and flavor, needs little introduction to lovers of Italian cuisine; it is the perfect accompaniment to tomatoes and pasta. Basil is a native of India, South Asia, and the Middle East, and it has also been grown throughout the entire Mediterranean region for centuries. It can flourish in temperate climates, but it should be grown either indoors in pots or out in the garden under cloches (clear plastic or glass coverings) until all danger of frosts has passed.

BASIL CAN
HELP TREAT

- *Anxiety and tension*
- *Congestion, coughs, and colds*
- *Colic*
- *Constipation*
- *Coughs*
- *Cuts and abrasions*
- *Diarrhea*
- *Flatulence*
- *Headaches and migraine*
- *Indigestion*
- *Insect bites and stings*
- *Muscle tension*
- *Nerve pain*
- *Sinusitis*
- *Sore throats*
- *Tiredness and lethargy*

Basil was introduced to Europe from India where basil leaves were often placed in the hands of the dead to ensure a safe journey to the next world. Similarly, since ancient times in Egypt and Greece, basil has been associated with death and it was believed to have the power to open the gates of heaven.

INTERNAL USE

As a medicine, basil has long been valued for its tranquillizing properties and its calming effect on the digestive system. It has been favored as a remedy to clear nervous headaches and congestion when taken as a snuff. In Japan, it has long been considered a useful herb for treating the common cold and, in Jewish lore, basil was said to lend strength while fasting, even when simply held in the hand.

Today, basil taken in food or made into a medicine makes an excellent digestive, enhancing appetite and promoting digestion and absorption. Its flavorful volatile oils have a relaxing effect throughout the digestive tract, relieving tension and spasm, gas and distension, nausea and indigestion, and diarrhea as well as constipation. Basil also has a relaxing effect on the nervous system, and makes an excellent natural tranquillizer. It can be used not only for indigestion and colic but

also for tight coughs, asthma, nervous headaches, migraine, muscle tension, and nerve pain. It can help to relieve anxiety and tension, lift the spirits, and enhance energy, and it makes a good remedy for clearing the mind and improving memory and concentration. Basil also has a decongestant action, particularly when taken in a hot infusion, helping to clear a stuffy head, colds, sinus and bronchial congestion. Its antiseptic properties enhance the body's fight against infection, whether in the digestive or respiratory systems.

EXTERNAL USE

Crushed fresh basil leaves can be rubbed on to cuts and abrasions, insect bites and stings to aid healing, and basil tea can be used as a steam inhalation for congestion.

In colder climates, basil grown outdoors should be protected by cloches.

HOW TO GROW

Basil is an annual herb. Propagate by sowing seeds under glass or indoors in spring. Transplant seedlings to a bed or to outside pots once all danger of frost has passed, planting them 1 ft (30 cm) apart. Basil does best in rich, well-drained soil in a warm, sheltered position, grows up to 3 ft (90 cm) and flowers in mid to late summer. Remove the flowers to stimulate bushy leaf growth and harvest from mid summer.

Marjoram

Origanum majorana

No herb garden would be complete without at least one of the members of the deliciously aromatic marjoram family. Sweet, or knotted, marjoram (*Origanum majorana*), with its white flowers growing in bundles, or knots, up the stem, is the sweetest smelling of all the marjorams, and it bears attractive gray-green leaves. As an alternative, wild marjoram, or oregano (*Origanum vulgare*), has a stronger flavor than sweet marjoram, especially when it is found growing on mountainsides in warm regions, such as the Mediterranean, the Middle East, and Asia.

MARJORAM CAN HELP TREAT

- *Anxiety and tension*
- *Chilblains*
- *Congestion, coughs, colds, and chest infections*
- *Colic*
- *Cramp*
- *Fevers and flu*
- *Headaches*
- *Insomnia*
- *Muscle tension and pain*
- *Period pain*
- *Poor circulation*
- *Stomach and bowel infections*
- *Urinary infections*

All members of the marjoram family of herbs can be used to enhance general health and promote a sense of wellbeing. Their warming and relaxing properties can be felt throughout the entire body.

INTERNAL USE

Marjoram improves circulation and relieves problems such as chilblains and cramps, and releasing the tension in muscles that is responsible for such problems as abdominal pain, menstrual cramps, headaches, and aching muscles.

Marjoram makes a good tonic for the nervous system, helping to reduce tension and anxiety, lift the spirits, improve energy levels, and yet induce restful sleep. It is an excellent remedy for all stress-related symptoms, particularly those associated with the digestive tract. It stimulates the appetite and promotes the digestion and absorption of food. Marjoram can also be taken to relieve indigestion, nausea, gas, spastic colon, and constipation.

The volatile oils in marjoram that lend the plant its spicy, aromatic taste and smell are antiseptic in nature and so make it a good medicine for stomach and bowel infections and for a wide range of other infections, whether bacterial, viral, or fungal. Marjoram is worth taking with or after antibiotics, to help to re-establish a normal bacterial population of the gut, and it makes an effective remedy to help ward off coughs, colds, flu, and fevers. If taken as a hot tea, marjoram reduces fevers and acts as an effective decongestant for treating colds, coughs, bronchial and nasal congestion, sinusitis, and hay fever.

The antioxidants found in marjoram assist in protecting the body against the ravages of the aging process, while its diuretic properties help to relieve fluid retention and promote the elimination of toxins from the system.

EXTERNAL USE

When added to rubbing oils, marjoram may be used to help to relieve stiff, aching muscles and painful arthritic joints.

HOW TO GROW

Sweet marjoram is usually grown as a half-hardy annual (one that is unable to withstand severe winter frosts) in temperate areas, but it can be grown as a perennial in warmer climates. Propagate marjoram by sowing seeds under glass and plant the seedlings out in spring, or once all danger of frost has passed. Space plants about 10 in (25 cm) apart. As an alternative, you can take stem cuttings in summer or divide the rootball in autumn and overwinter in a frost-free area. Marjoram grows best in light, well-drained soil and if it is planted in full sun it can reach a height of up to 2 ft (60 cm).

Parsley *Petroselinum crispum*

Parsley is the most popular of all culinary herbs, not only because of its delicious refreshing taste but also for its health-giving properties. Its popularity is nothing new, for it was valued by the ancient Egyptians, who used it as a remedy for urinary problems, and it was still in favor in the 16th century, when Culpeper wrote: "this herbe is so well known it needs no description." The great respect our ancestors had for parsley is understandable. It is highly nutritious, rich in vitamins A, B, and C, and in minerals, including iron, calcium, magnesium, manganese, and sodium, as well as in essential fatty acids. To gain most benefit from its nutrients, parsley is best eaten fresh.

PARSLEY CAN HELP TREAT

- *Anemia*
- *Anxiety*
- *Arthritis*
- *Bladder irritation*
- *Bruises and sprains*
- *Colic*
- *Flatulence*
- *Fluid retention*
- *Gout*
- *Headaches*
- *Insect bites and stings*
- *Period pain*
- *Poor circulation*
- *Tiredness and lethargy*
- *Vitamin and mineral deficiency*

The Greek physician Galen wrote "there is no herb so commonly used at table" and the Roman natural historian Pliny the Younger recorded that "parsley is in great request." Parsley certainly stimulates the appetite and promotes digestion and absorption, particularly of proteins.

INTERNAL USE

Parsley makes an excellent nutritious tonic when feeling tired and run down, when convalescing or anemic, because its digestive properties enable those with weak digestions to benefit from its health-giving vitamins and minerals. The volatile oil that gives parsley its characteristic taste and smell has a relaxing effect throughout the body, helping to ease spasm, colic, gas, and nervous indigestion as well as headaches, migraine, asthma, and an irritable bladder. In addition, parsley's volatile oils, especially apiol (*see* glossary), are thought to have antiseptic properties that help to combat infections.

Parsley has significance for the urinary system. The root, especially, stimulates the kidneys and has a diuretic action, helping to relieve urinary infections and fluid retention, and helping to clear from the system toxins which contribute to arthritis and gout. Parsley stimulates the uterus and can be used to promote menstruation, to relieve period pain, and to promote contractions during labor. Parsley also helps to stimulate the circulation and acts as a warming tonic. It has a supportive action in the nervous system, and can often help to relieve anxiety and mild depression.

EXTERNAL USE

Crushed fresh leaves can relieve the irritation caused by insect stings. Parsley leaves cooked in wine and applied as a poultice, is an old remedy for bruises and sprains. Parsley juice applied on cotton wool relieves toothache or earache.

Caution: Parsley should be avoided by sufferers of kidney disease, and during pregnancy since it may induce contractions.

HOW TO GROW

Usually treated as an annual, parsley can be slow to germinate, taking 4–6 weeks, so it is a good idea to soak the seeds for 24 hours in warm water before sowing them under glass or outdoors. Sow in early spring for summer use and again in midsummer for winter use. Plants should be thinned to 6–8 in (15–20 cm) apart once they are large enough to handle. Parsley prefers moderately rich, well-drained soil, in sun or partial shade. It grows 14–20 in (36–50 cm) high and bears flowers in mid to late summer. Leaves can be harvested when plants are 6 in (15 cm) tall.

Rosemary
Rosmarinus officinalis

This culinary herb is a native of Mediterranean shores, where its aroma can often be smelled on the warm sea air. Since the times of the ancient Egyptians and Greeks, rosemary has symbolized love and loyalty, friendship, and remembrance, and has played a part in rituals and ceremonies associated with both marriage and death. Rosemary has long been revered for its strengthening and tonic properties, in particular for the heart, brain, and nervous system. This may be explained by the fact that rosemary improves blood flow by stimulating the circulation.

ROSEMARY CAN HELP TREAT

- *Anxiety and tension*
- *Arthritis*
- *Asthma*
- *Congestion, coughs, and colds*
- *Chilblains*
- *Fevers and flu*
- *Fluid retention*
- *Hangovers*
- *Headaches and migraine*
- *Indigestion*
- *Infections*
- *Minor burns and scalds*
- *Poor appetite*
- *Poor circulation*
- *Sinusitis*
- *Tiredness and lethargy*

INTERNAL USE

Rosemary has a relaxing effect on the nervous system and a stimulating one at the same time, enhancing energy and concentration. It is an excellent herb to take in the morning.

Rosemary was a favorite in old apothecaries' shops for curing hangovers. By increasing blood flow to the head and releasing tension in muscles, rosemary is a useful herb for headaches and migraines when taken on a regular basis. Its warming and stimulating properties increase the flow of digestive juices and of bile from the liver, helping to improve digestion and liver function and thereby helping to detoxify the system—an additional benefit when treating migraines.

Rosemary's diuretic properties enhance the elimination of toxins via the urinary system. Modern research has shown that rosemary contains antioxidant substances that may slow down the aging process by inhibiting degenerative diseases.

Hot rosemary tea makes an excellent remedy to take at the first sign of colds, flu, and coughs and chest infections, and to help bring down fevers. The pungent and stimulating properties have an excellent decongestant action which can be put to good use in helping to treat not only congestion but also asthma, since rosemary's relaxant effect helps to relieve spasm in the bronchial tubes.

EXTERNAL USE

Rosemary oil, diluted in a base oil and rubbed on to the skin, has an invigorating effect, and by bringing blood to the surface it helps to reduce inflammation and speed healing. It can be used to heal cuts, sores, chilblains, and minor burns. The dilute rosemary oil makes an excellent remedy for soothing arthritic joints and aching muscles, and for treating local infections such as thrush. Rubbing rosemary oil on the temples or adding just a few drops to the bath makes an excellent "pick me up" to dispel lethargy and drowsiness and is recommended for those wishing to enhance concentration.

HOW TO GROW

Rosemary is a hardy, evergreen perennial in mild regions of Europe and North America and may be summered outdoors and wintered indoors in harsher regions. Propagate by sowing seeds in early spring, pricking out individual seedlings into pots when they are large enough to handle. They should be planted about 3 ft (1 m) apart. Alternatively, take 6–8 in (15–20 cm) hardwood cuttings from mid to late summer and plant in autumn. Rosemary does best in a slightly alkaline, well-drained soil, full sun and a sheltered position, since some forms are slightly tender. Rosemary grows up to 7 ft (2 m) high in a warm climate. It can be harvested all year around.

Sorrel *Rumex acetosa*

The young leaves of sorrel, a relative of rhubarb and dock, have a tangy, refreshing taste and make a piquant addition to salads. Sorrel soup is delicate and delicious, and puréed cooked sorrel makes a fine sauce for eating with fish. The name sorrel derives from the old French word *surele*, meaning sour, and the French variety, *R. scutatus*, or buckler-leaf sorrel, has long been popular in French cuisine; it is slightly less sharp than the garden sorrel, *R. acetosa*. French sorrel is native to the mountains of central and southern Europe, Turkey, and northern Iran, while garden sorrel is the cultivated form of the wild variety, with smaller leaves, that is native to Britain and Europe.

SORREL CAN
HELP TREAT

- *Acne*
- *Anemia*
- *Boils and abscesses*
- *Eczema*
- *Fluid retention*
- *Skin problems*
- *Vitamin C deficiency*

In 1629, herbalist John Parkinson wrote: "sorrel is much used in sauces both for the whole and the sicke . . . procuring unto them an appetite unto meat when their spirits are almost spent . . . and is also of a pleasant relish for the whole in quickening a dull stomacke that is overloden with every daies plenty of dishes."

INTERNAL USE

When eaten before a meal, sorrel leaves stimulate the appetite and aid the digestion of the food that is to follow. They have a laxative effect and a cooling action throughout the digestive tract. These cooling properties were put to good effect in the past as a remedy for fevers and for "cooling the blood." They have detoxifying properties, mainly through their diuretic action, aiding the elimination of toxins and wastes via the urinary system and thereby helping to cleanse the blood and remedy skin problems and other toxic conditions. Sorrel leaves are a cleansing remedy, like nettles and dandelions, helping to invigorate the system after the sedentary habits and heavy food of winter. A decoction of sorrel leaves is a remedy for cooling such skin problems as eczema, acne, boils and abscesses. A hot sorrel poultice can be applied to help heal abscesses

Due to its high vitamin C content, sorrel was once a popular remedy for scurvy, which is caused by a deficiency of vitamin C. It is also used for anemia since it is rich in iron. Sorrel is also a good source of carotenoids, chlorophyll, potassium, and magnesium. The carotenoids and vitamin C are natural antioxidants, which help to prevent damage caused by free radicals, enhancing immunity and helping to protect against degenerative diseases, heart and arterial disease, and possibly cancer.

Caution: Sorrel is rich in oxalates and should not be eaten regularly by those suffering from kidney disease or stones, arthritis, gout, or irritation of the stomach.

HOW TO GROW

Sorrel is a perennial which can be propagated by sowing seeds *in situ* in spring. Germination takes approximately 10 days, and seedlings can be thinned to 10 in (25 cm) apart when they are large enough to handle. Sorrel should be divided and replanted approximately every 5 years. Keep sorrel watered in dry weather and remove any of the flower heads that appear. Leaves can be taken from established plants from spring through to autumn, but choose small ones, since the larger leaves may have a very bitter taste. Dried sorrel has little flavor, so freezing is probably the best way to preserve the leaves.

Sage *Salvia officinalis*

Sage is a handsome shrub with highly aromatic velvety leaves and whorls of violet blue flowers. The ancient Greeks called it "the immortality herb" because it was believed to increase longevity. Sage has long been popular as a culinary herb, famous for the taste it imparts to sage and onion stuffing, and for easing digestion when eaten with rich and fatty foods such as goose. This traditional culinary use can be explained by its first-rate digestive properties.

SAGE CAN
HELP TREAT

- *Bronchial congestion*
- *Colic, cramping*
- *Cuts and abrasions*
- *Flatulence*
- *Fluid retention*
- *Gingivitis*
- *Gout*
- *Hot flushes and night sweats*
- *Indigestion*
- *Minor burns and scalds*
- *Mouth ulcers*
- *Nausea*
- *Period pain*
- *Poor concentration*
- *Respiratory infections*
- *Sore throats and tonsillitis*
- *Urinary infections*
- *Vaginal infections*
- *Vomiting*

Modern research has discovered the presence of antioxidant substances in sage, which may explain its beneficial action on the nervous system and its ancient reputation as a brain tonic and longevity herb. By helping to slow the effects of aging, and by its beneficial effect on the nerves, sage makes an excellent remedy to improve memory and alertness.

INTERNAL USE

Sage can help to treat indigestion, nausea, gas, bad breath, and excessive salivation. Its antispasmodic properties relax muscle tension throughout the body, and in the digestive tract this eases colic, stomachache, gas, and constipation. The astringent tannins in sage protect the gut lining from inflammation and help to relieve diarrhea.

Sage's antiseptic properties make it an excellent remedy for a wide range of respiratory infections—for sore throats, colds, flu, coughs, tonsillitis, and bronchitis—and a good decongestant at the same time, dispersing nasal and bronchial congestion accompanying coughs and asthma. It used to be a popular remedy for TB and other debilitating infections accompanied by profuse sweating and night sweats, since it can help to reduce secretions, such as mucus and sweat.

Its relaxing properties can be helpful during childbirth and to expel the placenta. Sage's

Cuttings of sage can be taken between mid and late summer. Rooting time is approximately 1 month in summer.

hormone-balancing and diuretic properties are particularly useful during the menopause for hot flushes and night sweats.

Sage makes an excellent cleansing remedy. Its bitter constituents enhance liver function, while the diuretic action aids elimination of toxins via the urinary system.

EXTERNAL USE

Sage's antiseptic and astringent properties can be put to good use as first aid for cuts, scalds and mild burns, sores and sunburn. Sage also makes an excellent gargle for sore throats, and a mouthwash for ulcers and inflamed gums. As well, sage can be used as a douche for vaginal infections such as thrush.

Caution: Sage is not recommended during pregnancy (*see above*), or when breastfeeding, as it can inhibit lactation.

HOW TO GROW

Sage is an evergreen perennial shrub which can be propagated by sowing seeds in trays in early spring. Prick out the seedlings when the first leaves show and transplant them in late spring. Sage grows best in well-drained soil and a sunny sheltered position. It grows up to 30 in (75 cm) tall and flowers in early summer. It can be harvested all year around.

Dandelion

Taraxacum officinale

This familiar plant with its bright yellow flowers is often considered a weed when it is found growing in lawns, flower borders, and meadows. In Europe it commands a deserved respect and is popular as a spring vegetable. A variety of cultivars are grown in gardens and blanched like endive to create more tender and less bitter leaves. Wild dandelions are found all over the world, comprising about 600 species; they are incredibly resilient plants, partly because they produce so many winged seedheads that they can reproduce without fertilization.

DANDELION CAN HELP TREAT

- *Boils and abscesses*
- *Constipation*
- *Fluid retention*
- *Gall bladder infections*
- *Kidney stones*
- *Liver problems*
- *Mastitis*
- *Poor appetite*
- *Prostate problems*
- *Skin problems*
- *Tiredness and irritability*
- *Urinary infections*
- *Vitamin and mineral deficiency*
- *Warts*

INTERNAL USE

Young dandelion leaves are highly nutritious, rich in antioxidant vitamins A and C, as well as vitamin B, potassium, and iron, while the long milky taproots have long been roasted and ground to make a pleasant caffeine-free coffee substitute. Young dandelion leaves are delicious in salads, and when included in cooked dishes they can be mixed with an equal quantity of spinach leaves to improve their bitter taste.

The French name for dandelion, *pissenlit*, and the old English name, piss-a-bed, inform us in no uncertain terms of the dandelion's renown as a diuretic. The leaves are particularly effective, and make the dandelion an excellent detoxifying remedy, useful for fluid retention, urinary infections, and prostate problems. A decoction of both the root and the leaves is an old folk remedy for dissolving kidney stones and gravel.

The bitter taste of the leaves and roots stimulate the bitter receptors in the mouth, which in turn send signals to the rest of the digestive tract and the liver to secrete digestive enzymes and bile. This stimulates the appetite and enhances the digestion and absorption of food. By helping to increase the flow of bile from the liver, the dandelion cleanses it. The young leaves have been eaten in spring for centuries as a cleanser, and a decoction of the root is sometimes used by herbalists to treat liver problems, hepatitis, gall

bladder infections, constipation, and for symptoms associated with a sluggish liver, such as tiredness, headaches, irritability, and skin problems. In China, dandelion root is used specifically to treat mastitis.

EXTERNAL USE

The milky juice of the dandelion stalks can be applied daily over several weeks to cure warts. Applying the leaves as a poultice is a traditional Chinese remedy for treating boils and abscesses. Tea made from the leaves and flowers has long been used as a wash for skin problems such as eczema and acne.

Caution: If sucked excessively by children, the milky juice from the dandelion stalk may cause nausea, vomiting, or diarrhea. Dandelions grown on a lawn that has been chemically treated should not be eaten.

HOW TO GROW

Dandelion will seed itself easily in the garden, but it can be propagated by sowing seeds in spring. The plants prefer full sun, but they will grow in almost any soil. They can grow up to about 1 ft (30 cm) high and flower from late spring to early autumn. Harvest the young leaves before they grow too large and bitter. Roots are best dug up in the autumn.

Thyme *Thymus vulgaris*

Thyme is native to the western Mediterranean and southern Italian regions, where the smell of wild thyme growing in the heat of the sun is wonderfully pungent. It is the essential oil in thyme that gives it this smell and its rich taste. For years, the main constituent of thyme's essential oil, thymol, has been used as an ingredient in antiseptic throat lozenges, cough remedies, vapor rubs, and mouthwashes.

THYME CAN HELP TREAT

- *Anxiety and tension*
- *Arthritis*
- *Asthma*
- *Bowel infections*
- *Congestion, colds, coughs, and chest infections*
- *Croup*
- *Cystitis*
- *Diarrhea*
- *Fevers and flu*
- *Gastroenteritis*
- *Gingivitis*
- *Irritable bladder*
- *Low immunity*
- *Mouth ulcers*
- *Poor circulation*
- *Sinusitis*
- *Sore throats and tonsillitis*
- *Thrush*

INTERNAL USE

Thyme's antiseptic properties can be put to good use in treating all sorts of infections: coughs, colds, sore throats, tonsillitis, flu, chest infections, and gastroenteritis, for example. Taken as a tea, a tincture, or an inhalation of the essential oil, thyme enhances immunity, and aids the body's fight against infection. Its expectorant action, combined with its antispasmodic effect, is excellent for moving phlegm out of the chest and easing tight, irritating, and hacking coughs.

In the digestive tract, thyme's antispasmodic action releases tension and can be used for treating irritable bowel syndrome and spastic colon. The tannins in thyme have an astringent action which helps to dry up secretions and to protect mucous membranes throughout the body from irritation, inflammation, and infection. The astringent action, combined with thyme's antiseptic properties, makes it worth using to treat diarrhea and bowel infections, and to re-establish a normal bacterial population in the gut after taking antibiotics or in the treatment of systemic candidiasis.

Thyme is a particularly effective remedy for those suffering with poor circulation, and hot thyme tea is an excellent revitalizing tonic when taken on a cold day. Recent research has discovered that the volatile oils in thyme have antioxidant properties, protecting the body against damage from free radicals and the onset of degenerative diseases such as arthritis and possibly cancer.

Thyme may help to relieve arthritis by other means as well. It has a generally cleansing effect, since it aids the elimination of toxins via the lungs, bowel, skin, and urinary system. Its diuretic and antiseptic properties are useful for treating cystitis when combined with soothing herbs such as marshmallow (*Althaea officinalis*) and for aiding the elimination of uric acid and toxins that contribute to arthritis.

EXTERNAL USE

Thyme's warming and stimulating effects can also be put to good use in liniments for arthritis and muscle pain. As an antiseptic, it can be applied in lotions to cuts, abrasions, and infections, used in gargles for sore throats, in mouthwashes for ulcers and bleeding gums, and in douches for vaginal infections such as thrush. Using thyme oil in a chest rub will help to clear coughs, colds, and sinusitis.

If cuttings are difficult to root, try layering the plant to encourage new sections to root. You can do this by pegging down one stem so that its underside is touching the soil.

HOW TO GROW

Sow seeds in pots in spring and transplant in early autumn. Cuttings can be taken in early summer and rooted in a cold frame. Thyme prefers well-drained soil and a sunny position and grows prostrate or upright to 1 ft (30 cm). Harvest leaves as needed and clip back after flowering and in autumn to encourage bushy growth.

This chapter outlines the principles and processes of organic growing.

the natural kitchen garden

Working with, not against, nature to develop a truly "green" garden isn't all digging and weeding— a union of traditional husbandry and modern technology helps to remove the drudgery and allows gardeners to see their seedlings grow and fruit, knowing that they are producing tasty, chemical- free foods for themselves and their families.

Organic gardening

Growing a kitchen garden is a way to establish a natural connection to the land, and organic gardening makes this connection a positive, creative one. For example, natural pesticides can be made from garlic, onion, seaweed, and even from slugs, which can be killed by leaving out a saucer of beer for them to drown in. Crop rotation and companion planting can make the soil rich and productive. Tough perennial weeds can be controlled by covering a plot with mulch.

Organic gardening allows you to feed the land that is feeding you. Vegetable peelings, tea leaves, coffee grounds, mown grass, and more can all be composted to make rich fertilizer. Healthy gardens produce robust vegetables, fruits, and herbs that, in turn, help make people healthy.

PLANNING YOUR GARDEN

Before planting your garden, it is a good idea to sit down with a pencil and paper and make a rough plan of your planting design. Every garden is unique, and planning a successful bed or border is not just a matter of personal preferences —your plan must also take into account the type of soil, the aspect of the site, shade from neighboring buildings and trees, the space available, and how much of it you wish to devote to fruits, vegetables, and herbs and how much to ornamentals. Planting plans have been included throughout this section for guidance.

SOIL

Healthy soil is essential for productive growth. Soil is a living entity, made up of sand, stones, clay, nutrients, organic matter, minerals and trace elements, as well as a wealth of microorganisms that contribute to its fertility. Anything that destroys these microorganisms, such as chemical fertilizers and pesticides, renders the soil less fertile.

Healthy soil

The main key to maintaining soil fertility and structure is a good supply of humus. This is decaying vegetable matter provided by plants that have died, and adds nutrients and bulk to the soil. It can also be applied manually in the form of compost or manure. Humus in the soil encourages the activity of bacteria, fungi, and the other microorganisms and invertebrates that live there. The action of these microorganisms releases the nutrients from the decomposing matter in the earth, which then dissolve in the soil water and are absorbed by the tiny hairs at the root tips of growing plants.

The best way to keep soil healthy is by continually keeping plants growing in it. The roots of the plants keep the soil broken up, allowing air and moisture to circulate freely, which is vital for the supply of nutrients to the roots. If left fallow, or uncultivated, clay soils will dry hard or become muddy, while sandy soils will either dry out, with a consequent danger of soil erosion, or rain will leach nutrients from the soil. It is important to avoid walking or driving equipment on the soil as much as possible; compacted soil will

PATIO HERB AND SALAD GARDEN

1 Dwarf nasturtiums
2 Garlic
3 Lettuces
4 Radishes
5 Chives
6 Calendulas
7 Arugula
8 Spring onions
9 Cilantro
10 Brick or paving stone paths
11 Dill
12 Carrots
13 Peas
14 Basil
15 Parsley
16 Zucchinis or cucumbers
17 Tomatoes
18 Thymes

form a crust, preventing oxygen from reaching the roots of the plants.

Heavy clay soils are usually rich in nutrients but need the addition of humus to improve their structure, to aerate the soil, to make the nutrient-rich water more readily available to plants, and to encourage good drainage. Leafmold, peat moss, or sharp sand will all help to improve the structure. More nutrient-rich compost and well-rotted manure can also be used. Light, sandy soils tend to lack nutrients, which are easily leached out by rain. The addition of humus in the form of compost or manure (*see pages 127–30*) will add nutrients and increase water retention.

UNDERSTANDING YOUR SOIL
Soil types

To get the best from your garden, it is important to get to know what kind of soil you have. As you get to know your soil, you will be able to feed and plant it appropriately. There are four main types:
1 *Clay soils* are heavy, rich in nutrients that are locked up because of a lack of air and freely moving water. Plant roots find it difficult to penetrate. It takes a long time to heat up in the spring and has poor drainage. It is very difficult to dig both when it is wet and when it is hard and dry (and best not attempted). Clay soil needs plenty of humus and some sand to improve its structure and release its nutrients. Humus can be dug into the soil or added in the form of an organic mulch—a layer of garden compost or other organic matter spread over the soil surface (*see page 125*).

Clay soil tends to be acid (*see page 124*) and, if so, will benefit from a light spread of lime over the surface in autumn. Never apply lime at the same time as manure because they react chemically and disperse the nitrogen. Lime should not be dug in, and plants should not be planted directly into soil that has just been spread with lime. Heavy soils

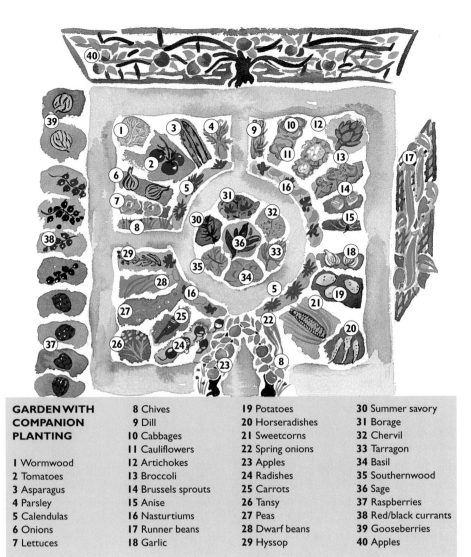

GARDEN WITH COMPANION PLANTING			
1 Wormwood	**8** Chives	**19** Potatoes	**30** Summer savory
2 Tomatoes	**9** Dill	**20** Horseradishes	**31** Borage
3 Asparagus	**10** Cabbages	**21** Sweetcorns	**32** Chervil
4 Parsley	**11** Cauliflowers	**22** Spring onions	**33** Tarragon
5 Calendulas	**12** Artichokes	**23** Apples	**34** Basil
6 Onions	**13** Broccoli	**24** Radishes	**35** Southernwood
7 Lettuces	**14** Brussels sprouts	**25** Carrots	**36** Sage
	15 Anise	**26** Tansy	**37** Raspberries
	16 Nasturtiums	**27** Peas	**38** Red/black currants
	17 Runner beans	**28** Dwarf beans	**39** Gooseberries
	18 Garlic	**29** Hyssop	**40** Apples

(when cultivated) are suitable for plants that need rich soil: leafy crops including beans, brassicas, peas, and potatoes.
To prepare: Dig or till in late autumn or early winter so that the frost can break up the lumps over the winter. Give it plenty of humus to lighten it and allow air in; the addition of garden compost, manure, leafmold, or sharp sand will, over several years, improve drainage.
2 *Sandy soils* are easy to work with and warm up quickly in the spring, but they dry out easily and tend to be deficient in nutrients, particularly potassium, which are easily washed out of the soil by rain. They also tend to be slightly acid and require humus in order to help retain water; compost and well-rotted manure will also add nutrients to the soil.
To prepare: If the soil is too acid, add lime in autumn, as for clay soils. Dig or till in the early spring, adding compost and/or well-rotted manure at the same time. Once the ground is planted, mulch all crops well to preserve moisture.
3 *Chalk soils* are alkaline. They tend to dry out easily and benefit from mulching (apply over well-watered ground). They are fairly fertile but may lack nitrogen.
To prepare: Dig or till in compost or well-rotted manure and peat moss in early spring. To increase the nitrogen content, dig in a "green manure" (*see pages 128–9*) of peas, annual lupines, or comfrey.

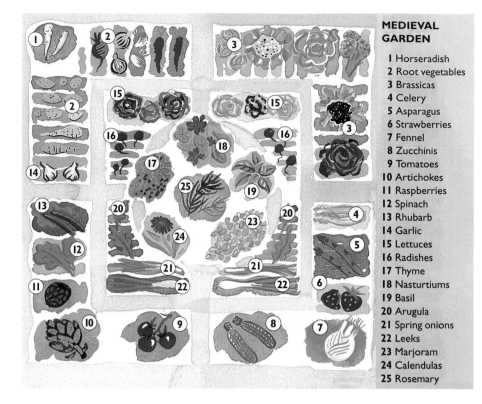

4 *Loam* is the type of soil that all gardeners aim to achieve—a well-balanced mixture of clay, silt, and sand with a good structure, high fertility and humus content, good water retention, and good drainage. It is easy to work, and perfect for most crops. Almost any soil can reach this ideal state if it is worked well and planted over a long period.

Testing for pH

As well as understanding what type of soil you have, you will also need to know whether your soil is acid, alkaline, or neutral, and this may vary from one part of the garden to another. A pH (potential of Hydrogen) test is available from all garden stores and is easy to use.

Soil acidity is measured from 0 to 14. An average-to-normal (neutral) soil has a pH of 7; a lower reading indicates an acid soil; higher an alkaline one. Acid soils will particularly suit potatoes, strawberries, and raspberries; slightly acid to slightly alkaline soils suit such plants as beets, broccoli, and cauliflower.

If your soil is very acid or alkaline, you may find it deficient in certain nutrients. Acid soils can be corrected by adding a top-dressing of lime. Alkaline soils can be harder to correct, but will certainly benefit from well-rotted manure. Additional lime should obviously be avoided.

Nutrients in the soil

Healthy plants need certain essential elements that should be naturally present in the soil. Any lack or imbalance of these can cause deficiency diseases. The three main requirements for healthy plants are nitrogen, phosphorus, and potassium. Nitrogen is required for healthy green leaves and stem growth; phosphorus for good root growth and for the production of fruits and seeds; and potassium for disease resistance and the maintenance of healthy growth.

Plants also require calcium, magnesium (needed in the production of chlorophyll), and sulfur, as well as trace elements (or micronutrients) such as manganese, iron, zinc, copper, and molybdenum, which exist naturally in the soil in minute quantities. They can also be obtained from compost and well-rotted manure.

Deficiencies in the soil

Deficiencies are usually caused by lack of good organic matter in the soil and the use of chemical fertilizers, which upsets the natural balance of minerals. A lack of nitrogen leads to stunted growth, while a lack of potassium leads to an increased susceptibility to disease as well as to discoloration of the edges of leaves.

In order to avoid or remedy deficiencies, it is important to adopt a good regime of composting and adding organic matter to the soil. In extreme cases, wood ash will supply potassium, and cottonseed meal, fishmeal, or fish emulsion will supply nitrogen. Excess amounts of potassium, calcium, or nitrogen bring their own problems as they create deficiencies in other nutrients, so it is important to aim for a good, healthy balance.

Plants may also fail to thrive because the soil you plant them in has the wrong pH. Blueberries like very acid soil (pH 4.5 to 5.2). Potatoes like acid, too, doing well in soil with a pH of 5.3 to 6.0. Strawberries, raspberries, and asparagus like it slightly acid, and broccoli prefers a pH of 6.4 to 7.0, slightly acid to neutral. Beets, on the other hand, like slightly alkaline soils with a pH of 7.5 to 7.8.

Too much water can be worse than too little, and some plants like full sun while others prefer partial shade. Pay attention to your plants' individual needs and they will do well for you.

Drainage

Poor drainage can be a problem on very wet sites and where there is heavy clay soil. This problem will be compounded if the soil is often trodden on, since this causes compaction. Signs indicating poor drainage include horsetails on neglected sites, puddles left after rain, and moss on lawns. Poor drainage is bad for plants because excess water deprives the roots of oxygen and the necessary bacteria cannot thrive. As a result, plants will fail to grow well, and may wither and die. (*See page 127 for information on how to improve soil drainage*).

IMPROVING YOUR SOIL
The purpose of digging

The general purpose behind digging or tilling is to improve the texture of the topsoil by breaking up heavy soils, improving aeration and drainage, adding humus in the form of compost and/or well-rotted manure, digging out perennial weeds, and digging in annual weeds before they run to seed. Digging or tilling is particularly necessary on poor soils, and on soil that is being prepared for cultivation for the first time.

It is essential when digging or tilling to avoid mixing the infertile subsoil with the fertile topsoil (the color of the subsoil will be noticeably lighter). Avoid digging or tilling when the soil is dry, since any moisture will be lost in the process. Dig or till heavy soils in autumn so that the frost can cause the water in the earth to freeze, expand, and thus break up the large clumps. Dig or till light soils in early spring, incorporating plenty of organic matter to give the soil "body."

The no-dig system

A debate rages about whether to dig or till at all, and there are a number of good reasons for not digging or tilling once the ground has been initially prepared. After one thorough double digging, with large amounts of compost or well-rotted manure dug in, further organic matter can be provided annually as a mulch or top-dressing. A mulch laid on the surface (*see below*) will be taken down into the soil by earthworms and decomposed by microorganisms to add humus to the soil and so increase its fertility; it will also protect the soil from drying out. A no-dig system is certainly better for light soils, where it is important to retain moisture, and it is obviously ideal for permanent crops (such as fruit bushes and asparagus) where the surrounding soil can be forked over and watered before a mulch is laid on the surface. It also obviates the necessity of walking or driving equipment on the soil and risking compaction.

Raised beds fulfill the same purpose and are very useful where there is poor drainage. The beds can be constructed so that even the center can be reached from an adjacent path: the maximum width should be 4 ft (1.2 m) and they should be 12–30 in (30–75 cm) high. The walls can be built of any suitable building material, such as untreated wood, brick, or stone. They must be well constructed so that they do not collapse if you lean on them, or through the weight of the soil they are retaining.

If drainage is a particularly bad problem on your plot, it is best to make the walls slightly higher and put a good layer of sand, gravel, and stones on the bottom. Top with good soil, lots of organic matter (compost or well-rotted manure), and sharp sand if this is needed to provide extra drainage.

Organic mulches

As an alternative to digging or tilling garden compost or very well-rotted manure into the soil in order to maintain its fertility, it can be laid on the surface in the form of a mulch to a depth of 2–4 in (5–10 cm). You can also use leafmold or

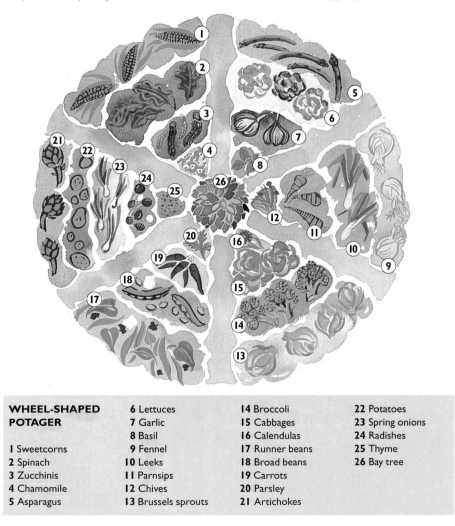

WHEEL-SHAPED POTAGER	6 Lettuces	14 Broccoli	22 Potatoes
	7 Garlic	15 Cabbages	23 Spring onions
	8 Basil	16 Calendulas	24 Radishes
1 Sweetcorns	9 Fennel	17 Runner beans	25 Thyme
2 Spinach	10 Leeks	18 Broad beans	26 Bay tree
3 Zucchinis	11 Parsnips	19 Carrots	
4 Chamomile	12 Chives	20 Parsley	
5 Asparagus	13 Brussels sprouts	21 Artichokes	

grass clippings to add humus to the soil (*see the following list*). This system relies on the activity of earthworms to do the digging for you.

The worms take the humus down into the topsoil and their little tunnels provide drainage and allow access for oxygen, while the worm casts provide extra nutrients. The ground should be wet before a mulch is applied, since mulching helps to conserves moisture. Mulching also keeps the soil warm, provides essential nutrients, and controls the growth of annual weeds by depriving them of light. Surface mulching will not, however, control the activity of perennial weeds, which need to be dealt with by hand, making sure that all roots are removed.

Good organic mulches:
- Garden compost (*see page 127*).
- Well-rotted farmyard manure (cow or horse mixed with straw or hay). Horse manure with wood shavings can also be used as a mulch, but must not be dug in, since the the shavings inhibit plant growth by using up the available nitrogen as they decompose.
- Leafmold, which is excellent for improving the soil texture but is deficient in nutrients.
- Composted straw and hay. Layers of uncomposted straw can also be used to protect plants from frost.
- Commercial wood bark.
- Seaweed, which is very rich in nutrients. This material can be collected from beaches where there is no pollution after it has been ripped up by the action of the sea.

Digging

Single digging or tilling to one spade's depth—or one "spit," which as about 1 ft (30 cm)—is done to incorporate garden compost or well-rotted manure in ground that has previously been cultivated.

Standard double digging or tilling to two spades' depth—or two "spits," which is about 2 ft (60 cm)—is best suited to previously uncultivated plots in order to break up compacted soil, improve drainage, remove deep roots of perennial weeds, and improve fertility by incorporating compost or well-rotted manure.

Single digging or tilling

1 Dig a trench one spit deep across the width of your plot, and reserve the soil in a wheelbarrow or on some plastic sheeting.

2 Scatter compost evenly in the bottom of the trench.

3 Dig an adjacent trench, filling the first trench with the soil from the second, forking it well to mix it with the compost. Do not leave any compost on the surface.

4 Scatter compost evenly in the bottom of the second trench.

5 Dig a third trench and repeat the process until the allotted space has all been composted. Fill the last trench with the soil reserved from the first trench.

Double digging or tilling

1 Dig a trench one spit deep across the width of your plot and reserve the soil in a wheelbarrow or on some plastic sheeting.

2 *Using a fork, break up the soil at the bottom of the trench to the depth of one spit.*

3 *Add a layer of compost evenly on top of the broken soil.*

4 *Dig an adjacent trench, using the soil from the second trench to fill the first.*

5 *Repeat the process until the allotted space is dug and composted. Fill the last trench with the soil reserved from the first trench.*

Double digging or tilling breaks up the soil to a deeper level and is, therefore, appropriate where deep-rooted crops are to be grown, but it is mainly useful because it provides a good preparation of uncultivated soil for planting.

However, some experts do not recommend double digging because there is a danger of the valuable topsoil, which contains most of the soil bacteria, earthworms, and organic matter, becoming mixed with the less fertile subsoil. Careful double digging or tilling, following the instructions shown above, ensures that this does not happen.

Preparing an uncultivated or neglected plot

When taking over a new patch, the soil will probably lack nutrients and a good structure. It will first need double digging, incorporating lots of garden compost and/or well-rotted manure. You should test for acidity and ascertain the type of soil in order to determine how best to improve it.

If there are lots of perennial weeds, covering the whole area with old carpet or thick plastic for one growing season will get rid of most things except some of the most tenacious weeds, which may need two growing seasons, since the

roots of some weeds can go down 2 ft (60 cm) or more. Otherwise, double dig or till (*see opposite and above*), removing as many roots as possible.

Improving drainage

To improve the drainage on an unprepared site, use standard double digging, adding sharp or coarse sand to the compost or well-rotted manure that is forked into the second spit. Drainage should gradually improve as organic matter is added to the soil over a period of time. If this has no effect and the soil remains waterlogged, there is a clear need to construct a drainage ditch or lay pipes. An easier alternative is to make raised beds (*see page 125*).

Compost

The organic matter provided by compost is an invaluable source of nutrients and humus for the soil and, hence, your fruit, vegetables, and herbs. It is also a marvelous means of recycling kitchen waste and all green material from the garden. Ingredients that are good to add to the compost heap include vegetable and fruit peelings from the kitchen, dead cut flowers from the house, tea leaves, coffee grounds, any thinnings and prunings, grass clippings, and annual weeds.

Grass clippings and greens serve a particular purpose in a compost heap: if the heap is well constructed (*see below*), the heat generated by the decomposition of the clippings and other green matter (caused by bacterial activity) will be high enough to kill the roots of perennial weeds and weed seeds. If you are not able to construct the compost heap in such a way that these temperatures are achieved—in the range of 150–158°F (65–70°C)—avoid putting weed roots and weed seeds onto the heap.

Concentrated poultry, goat, and rabbit manure, or chopped comfrey, can be added in layers to activate the compost.

It is best to avoid putting woody material on the compost heap. Instead, shred it and use it as a mulch for roses, strawberries, and raspberries. Don't put cooked food on the heap, since this will attract rats and other animals; or soil (shake it out of the roots of weeds), or sawdust. Discard any diseased plants.

How to construct a compost heap

There are several ways to achieve a well-built heap. You can construct or buy a wooden compost box, and a variety of good plastic compost bins is available. These have lids to retain moisture and heat, but are open at the bottom to allow air to circulate and worms to work their way up from the ground; their activity is important in the process. Since the worms also take the compost down into the soil, a portable bin has advantages. It can be moved annually and the nutrient-rich soil underneath planted up. Ideally, you should have two bins—one with compost ready for use, and the other left with new material to compost down. While the second one is in use, the first can be refilled and left to compost.

First, decide on the kind of bin you want. Whether you buy or make a bin, bear in mind that a good compost heap tends to be about 3 ft (90cm) across and the same deep. If it is smaller than this, it will not generate enough heat to kill weed seeds and harmful bacteria. At the bottom of the bin, make a layer of woody material such as cornstalks, twigs, or bush clippings. Over this make a second layer, 6–8 in (15–20 cm) thick, of green material—grass clippings, mixed with vegetable trimmings and green garden waste. Then put down a layer of brown material such as dead leaves. Ideally, chop up the dead leaves first with a shredder or even run over them with a lawnmower. Sprinkle on some mature compost or well-rotted manure, then repeat the green and brown layers until the pile is at least 3 ft (90 cm) deep.

Don't bother with so-called compost activators—all the microorganisms that are needed exist naturally in the material of the heap. Do not add lime, since this will upset the essential carbon-nitrogen balance of the heap. If you leave it undisturbed, the compost will be ready to use in 6 to 12 months. However, if you turn it frequently (once a week, for example), you can have usable compost in as little as 8 to 12 weeks.

Manures

Manures, another form of decomposed organic matter, also help to maintain the fertility of the soil, as well as providing

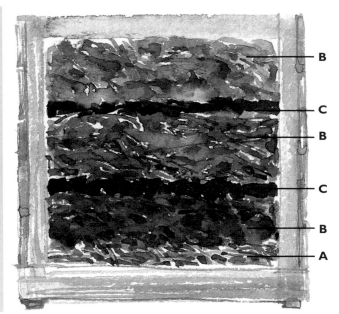

COMPOST HEAP

A Woody material such as twigs and bush clippings.

B Green material such as lawn mowings, green garden waste and vegetable trimmings.

C Brown material such as dead leaves.

humus. Manures will improve the condition of the soil, add nutrients, and help to protect against drought. By supplying decomposing organic matter they, as with compost, provide food for the microorganisms living in the soil, which, in turn, release nutrients for the plants. However, an excess of manure in the soil will cause it to become too acid—requiring an application of lime.

Green manures are quick-growing leafy crops, such as mustard, vetch, lupines, borage, and buckwheat, that can be grown over a short season and dug back into the soil to enrich it. They can also be used to help retain nutrients in fallow ground, which is often susceptible to leaching. The sappy stems and leaves break down quickly, adding extra humus and nutrients to the soil and encouraging the activity of earthworms. In warmer climates, such winter crops as winter rye will help to protect the soil so that it is not left open to the elements, when nutrients can be washed away.

All green manures will prevent loss of nitrogen from the soil, while decomposing lupines and vetches add valuable extra nitrogen thanks to the nitrogen-fixing bacteria in their roots. These crops must always be cut before they go to seed, however—usually at a height of about 8 in (20 cm). They are useful on areas that would otherwise be left bare after main food crops have been harvested. They will deter or even kill annual weeds by starving them of light.

Green manures for spring sowing

Mustard is very good for new plots, but it should not be grown year after year since it can encourage clubroot. Seeds should be sown thickly between midspring and midsummer, and plants dug in when they reach 12 in (30 cm) in height.

Mustard will deter wireworms (the larvae of click beetles), which attack the roots of carrots and potatoes, and will ward off snails. However, as with the cabbage, mustard is a crucifer and may increase the incidence of clubroot if this is already a problem.

Lupines are a good crop to cultivate before planting strawberries as their roots have phosphorus-gathering fungi. Sow seeds between midspring and midsummer. Dig in the plants before they flower and before they become too stemmy.

Vetches are members of the pea family and add valuable nitrogen to the soil.

Purple vetch (*Vicia atropurpurea*), common vetch (*V. sativa*), and hairy vetch (*V. villosa*) choke out weeds and feed beneficial insects, too. They yield substantial amounts of organic matter. Seeds should be sown in early autumn or spring.
Buckwheat (Fagopyrum esculentum) is principally used to attract bees and hoverflies, which then feed on aphids. Buckwheat is particularly suitable for heavy clay soils and has attractive white flowers, which look decorative in the garden. Sow seeds from midspring to midsummer.

Green manures for winter sowing

Winter rye (secale cereale) is a hardy winter crop that prevents soil erosion and, when dug in the following spring, adds organic matter to the soil. It can be sown anytime from spring to midautumn.

Animal manures

These contain feces, urine, and bedding materials from animals and poultry, and are rich in nitrogen and phosphorus. They also supply trace elements and bacteria, as well as humus, which helps to provide good soil texture.
Horse manure is usually available from a local stable or a friendly horse-owning neighbor, and consists of a mixture of horse dung and straw. It tends to be wet and heavy to move about but has the advantage of being mainly free from pesticides. If it is well rotted, it can be dug straight in, but it often requires further rotting down for at least a year. It should be stacked on top of a plastic sheet to collect the liquid that will drain out. Position it against a wall or mound it up, and compress the heap as much as possible by treading it down to exclude air. Hose it down if it is dry and cover it with plastic or a corrugated sheet in such a way that the rain will drain off, to prevent nutrients leaching out from the heap. Covering it will also prevent nitrogen leaking out in the form of gas and will trap in bad smells. If the manure is dung

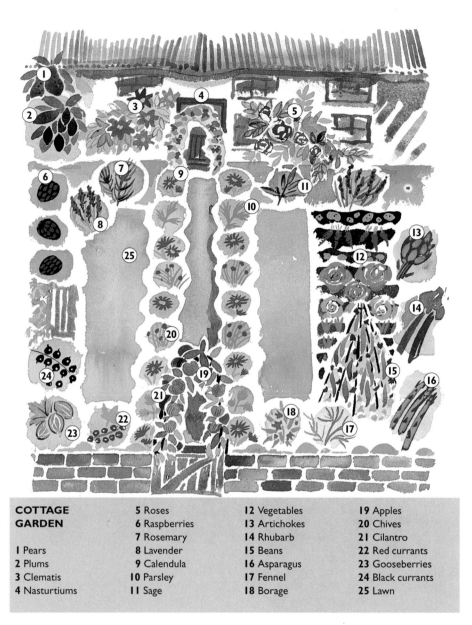

COTTAGE GARDEN			
1 Pears	5 Roses	12 Vegetables	19 Apples
2 Plums	6 Raspberries	13 Artichokes	20 Chives
3 Clematis	7 Rosemary	14 Rhubarb	21 Cilantro
4 Nasturtiums	8 Lavender	15 Beans	22 Red currants
	9 Calendula	16 Asparagus	23 Gooseberries
	10 Parsley	17 Fennel	24 Black currants
	11 Sage	18 Borage	25 Lawn

and wood shavings (not straw), this should not be dug in but it can be used as surface mulch once it is well rotted.
Poultry manure, if you can obtain it from a poultry breeder, usually comes in the form of deep litter incorporated with straw, and is rich in potassium, nitrogen, and phosphorus. It is normally dry and easy to transport. It is rich and concentrated, so you should only use a 2-gallon bucket per square yard (10-liter bucket per m^2), dug into the soil in autumn or spring. Poultry manure that is not from deep litter is even more concentrated and

should be used sparingly only as a top-dressing or as a compost heap activator. If possible, avoid poultry manure from factory farms because it may contain traces of antibiotics; free range is best.
Pigeon manure is usually dry and easy to transport. It should not be put directly on the garden, but used on the compost heap as an activator. Store it in a plastic (not metal) garbage can and sprinkle it on top of a layer of green kitchen waste or grass clippings (*see pages 127–8 for building a compost heap*). Avoid breathing the dust, since it can cause lung disease.

excellent mulch and also an environmentally friendly substitute for peat moss. It contains fewer nutrients than compost or manure but can be used as it is, or combined with grass clippings or with chopped comfrey leaves to increase the nutrient content of the soil.

Leafmold kept for one year is good enough to dig into soil, and after two years it can be used as potting mix. Oak and beech produce the best leaves for making leafmold. Avoid holly, laurel leaves, and pine needles, since they do not rot down well.

How to make leafmold

Leafmold is made through the action of fungi, which need no oxygen. To speed up the rotting process and enrich the compost you can mix it up with grass clippings, equal parts of chopped comfrey, or urine. Construct a frame for a leafmold heap with stout posts at the corners and wire netting to make the sides. The main object is to stop the leaves blowing about, though it also helps if they are kept damp. You can also make smaller quantities of leafmold in black plastic garbage bags, which can be kept in a shed or a corner of the garden (make sure that there are no weeds mixed in). This is the best method if you are making comfrey leafmold for potting mix (*see pages 145–6*), because it breaks down more quickly than if left in the open, and should be ready in a few months.

Peat substitutes

Peat moss is sterile (it contains no weed seeds), adds humus, helps to lighten heavy soils, and plant roots can readily find their way through it. It is also light and clean to transport, is a good base for seed and potting mixes, and it makes a useful medium for storing vegetables. However, peat is a finite resource and to continue to use it would cause irrevocable damage to the environments from which it is extracted. As a result of this

VERTICAL GARDEN	8 Tomatoes	17 Chamomile	27 Pears
	9 Red/Black currants	18 Rosemary	28 Plums
1 Raspberries	10 Calendula	19 Damsons	29 Rhubarb
2 Peas	11 Blackberries	20 Carrots	30 Artichokes
3 Nasturtiums	12 Thyme	21 Potatoes	31 Pumpkins/squash
4 Lettuces	13 Parsley	22 Dill	32 Strawberries
5 Radishes	14 Cilantro	23 Sage	33 Climbing squashes
6 Spring onions	15 Basil	24 Cherries	34 Runner beans
7 Arugula	16 Trailing nasturtiums	25 Marjoram	35 Ballerina apples
		26 Peaches	

Pig and cow manures, as with poultry manure, is best obtained from an organic farm, since factory farm manure may contain traces of antibiotics and pesticides. Farmers sometimes add copper sulfate to pig feed, which is toxic to the soil—so check this at the source. Stack it as you would horse manure.

Goat and rabbit manures are both good compost activators. Goat manure is rich in potassium and nitrogen; rabbit droppings in nitrogen and phosphorus.

Leafmold

Leafmold provides wonderful humus and generally improves soil fertility. It is an concern, much research has recently gone into finding good substitutes.

For improving the soil, you can substitute well-rotted manure, compost, or leafmold. You can also buy shredded bark or composted and dried animal manures. (*For alternatives to peat for use as a mulch, see pages 125–6, and for alternatives to peat for use as a seed or potting mix, see pages 145–6.*)

WEED CONTROL

PERENNIAL WEEDS

It is best to attempt to remove deep-rooted perennial weeds before planting, as once the garden is planted, the weeds will be far harder to get rid of, although continual cutting down will weaken and eventually kill them. Perennial weeds include couch grass, field bindweed, nettle, dock, and dandelion.

There are several ways of clearing persistent weeds from a new site:
1 You can double dig or till and remove all the roots you can find by hand. In fine weather, you can leave them on the surface of the soil to dry, then rake them up and discard them, or compost them if your compost heap gets hot enough.
2 You can rototill the site to chop up the weed roots. If you do this regularly over a period of time, most weeds will gradually become exhausted and die. However, the chopped roots of field bindweed and couchgrass will not die, and each piece will resprout.
3 You can cover the plot with old carpet or black plastic, making sure that it is well buried or held down with rocks at the edges to stop it lifting in the wind. This method kills weeds by light starvation. The covering is best put down in very early spring and left for one whole growing season. A thick mulch of newspaper between soft fruit bushes, held down with rocks or a 6 in (15 cm) layer of straw will deter couch grass.

CITY GARDEN

1 Peas
2 Bush basil
3 Peppers
4 Chilies
5 Tomatoes
6 Arugula
7 Radishes
8 Cilantro
9 Mint
10 Ballerina apple tree
11 Peach tree
12 Strawberries
13 Lettuces
14 Chives
15 Parsley

4 Nettles are weeds but they also make excellent fodder for the compost heap, where they work as an activator, and they can also be turned into an excellent liquid manure (*see page 136*). In addition, they provide a good habitat for caterpillars and aphids, which, in turn, feed ladybugs. For all these reasons it is worth keeping a corner of the garden for nettles—and, as a bonus, the young shoots make a delicious and nourishing soup.
5 Weeds with long, anchoring taproots, such as docks and dandelions, can be lifted out with a spade. The easiest time to do this is in early spring, when the soil is moist and when they have less hold on the soil. If the root breaks, it will sprout again. They will also be weakened by continual cutting back. Docks can be cut back in early summer and then again in early autumn. (You can put the leaves on the compost heap as long as there is no danger of them seeding.)

ANNUAL WEEDS

Annual weeds appear in succession from early spring. Learn to recognize them and deal with them as soon as they appear.

Groundsel, chickweed, and speedwell can all be hoed or hand-weeded. Meadow grass is best hand-weeded because it resprouts from the root.

GENERAL MAINTENANCE

• Mulching helps to prevent the growth of weeds by depriving them of light.
• Always deal with weeds before they flower and seed. Hand-weed or hoe them while they are still small, as they will be harder to deal with afterward.
• When preparing soil for planting, dig in annual weeds by turning them under the soil and burying them, or pull them and add them to the compost pile.
• Between shrubs, fork or hoe weeds into the ground.
• Between tender plants, vegetables, and herbs, hoe or hand-weed.
• Try to keep the soil in cultivation so that weeds have nowhere to grow.
• Hoe only $1/2$ in (1.5 cm) deep to prevent damage to roots of nearby plants.
• If possible, hoe in warm weather when the weeds will quickly wilt on the surface. Avoid hoeing if it looks as if it might rain.

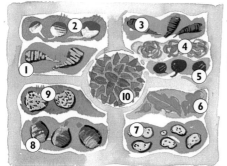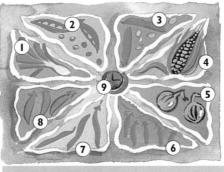

CROP ROTATION Bed 1

1 Parsnips	5 Beets	9 Celeriac
2 Turnips	6 Spinach	10 Bay tree
3 Carrots	7 Potatoes	
4 Lettuces	8 Rutabagas	

CROP ROTATION Bed 2

1 Leeks	5 Onions	8 Snap peas
2 Peas	6 Dwarf beans	9 Sun dial
3 Broad beans	7 Runner	
4 Sweetcorn	beans	

CROP ROTATION Bed 3

1 Red	3 Calabrese	6 Broccoli
cabbages	4 Green	
2 Brussels	cabbages	
sprouts	5 Cauliflowers	

CULTIVATION

CROP ROTATION

Crop rotation means growing different crops in different parts of the vegetable plot in annual succession. Crops benefit and deplete the soil in lots of ways, and crop rotation helps to prevent the soil becoming exhausted as well as discouraging disease.

Crop rotation depends on grouping vegetable crops as "families," according to their type and also to the demands they make on the soil. Different plant families require different elements from the soil, or the same elements but in different quantities. A crop rotation can be organized over a 3, 4, or even a 5-year period, depending on how many crops you want to rotate. A 3-year rotation is most common in small gardens, but a 4-year rotation is preferable if space allows.

Vegetables for rotation are categorized into root crops and tubers; brassicas; alliums; and legumes.

Roots and tubers are usually divided into early and main crop potatoes in one plot and all other roots (and tomatoes) in another, though it is perfectly possible to grow them in a single plot. Members of the allium family (onions, shallots, leeks, garlic) and summer squash (which is usually grown with the allium family) will be grown in another plot, brassicas in a third, and legumes in a fourth.

Crop rotation is a good, balanced organic approach to vegetable gardening. It prevents the depletion of the same nutrients year after year (which can result in a need for chemical fertilizers), and improves the general fertility of the soil. For example, potatoes will help to break up the ground on a new plot during the first year, and will grow well even in poor soil. These can be followed in the second year by legumes, which have nitrogen-fixing nodules on their roots. These legumes—cut down, not dug up after cropping, with the nitrogen-fixing roots left in the ground to rot, and extra compost or well-rotted manure added— prepare the way for the "hungry" brassicas in the third year. In a 4-year rotation, these can then be followed by a planting of other root crops that are less nitrogen-hungry. Lettuces and other quick-growing crops, such as salad onions, spinach, green beans, peas, and radishes, can be grown as follow-ons or fill-ins (catch-crops) after the main crop has been harvested so that land is not left fallow and is always under cultivation.

Crop rotation helps to control soil-borne pests and diseases, which tend to run in families. Cabbage family clubroot spores, for example, can live in the soil for 20 years, but even a 4-year rotation will help to check this disease. White rot (a fungus) found on onions and scab on potatoes are kept in check by crop rotation, which also helps to protect against eelworms on potatoes.

When using crop rotation you need to apply common sense regarding your particular plot. Plant what you enjoy eating and what grows well in the conditions that prevail in your garden, and choose disease-resistant varieties that are suited to your soil. You will also need to leave room for permanent crops, such as fruit trees or bushes, rhubarb, and asparagus.

All crops mature at different times of the year, so plan for a planting succession: spring cabbage, kale, and leeks can be put in after the main crop potatoes have been harvested in late summer; early carrots, leeks, and celery can be put in to follow the early potatoes in mid summer. Catch-crops, such as lettuces, salad onions, radishes, or spinach, can fill in anywhere.

Crop families

Brassicas: broccoli, brussels sprouts, cabbages, cauliflowers, kale, radishes, kohlrabi, rutabagas, and turnips.

Legumes: beans (broad, snap, haricot, scarlet runner) and peas.

Root crops: beets, carrots, celery, Jeru-

Jerusalem artichokes, parsnips, potatoes, and salsify.

Alliums: garlic, leeks, onions, and shallots.

Rotation plans

Three-year plan: The rotation is as follows: in the 2nd year, Bed 1 moves onto the plot previously occupied by 3, Bed 3 moves onto the plot previously occupied by 2, Bed 2 moves onto the plot previously occupied by 1, and so on (*see below*).

Bed 1 Root crops—such as beets, carrots, potatoes, parsnips, and Jerusalem artichokes—can be followed by spinach or lettuce to fill in

Bed 2 Legumes—such as runner beans, broad beans, snap beans, peas, and also sweetcorns, celery, onions, and leeks

Bed 3 Brassicas—such as brussels sprouts, cabbage, cauliflower, broccoli, and kohlrabi

Four-year plan: The rotation is as follows: in the 2nd year, Bed 1 moves onto the plot previously occupied by 4, Bed 4 moves onto the plot previously occupied by 3, Bed 3 moves onto the plot previously occupied by 2, Bed 2 moves onto the plot occupied by 1, and so on.

Plan A

Bed 1 Roots—celery, carrots, beets, parsnips, potatoes, and turnips—as well as peppers, tomatoes, and celeriac

Bed 2 Legumes

Bed 3 Brassicas

Bed 4 Leeks, onions, shallots, and garlic

Plan B

Bed 1 Early and main crop potatoes

Bed 2 Brassicas

Bed 3 Legumes

Bed 4 Roots—beets, carrots, parsnips, —Jerusalem artichokes, and salsify, as well as onions

Summer squash and pumpkin can either be planted with the onion family or with the root crops.

COMPANION PLANTING

Companion planting is a great addition to crop rotation. It is based on the simple principle that all living things are interdependent and that several plants are mutually beneficial to each other. This interrelationship can be beneficial in different ways: plants used for hedges provide warmth and shelter to others nearby; some plants attract pollinators and predators such as ladybugs and lacewings to the area; highly scented vegetables and herbs, and flowers rich in essential oils deter pests by confusing the scent. For example, onions planted near carrots deter carrot rust fly and French marigolds help to prevent eelworm on potatoes. Aromatic herbs such as lavender, tansy, and sage deter ants; rosemary, thyme, and peppermint deter cabbage butterflies and slugs—as do French marigold, garlic, and onions. Artemisias discourage moths and insects that attack brassicas and carrots, and rue's unpleasant smell keeps away most pests. Tall

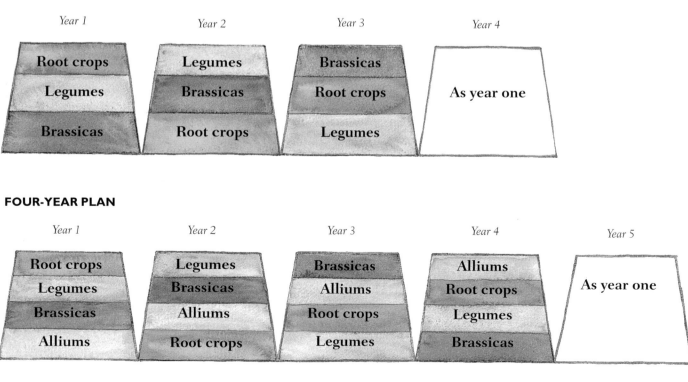

THREE-YEAR PLAN

Year 1 — Root crops / Legumes / Brassicas
Year 2 — Legumes / Brassicas / Root crops
Year 3 — Brassicas / Root crops / Legumes
Year 4 — As year one

FOUR-YEAR PLAN

Year 1 — Root crops / Legumes / Brassicas / Alliums
Year 2 — Legumes / Brassicas / Alliums / Root crops
Year 3 — Brassicas / Alliums / Root crops / Legumes
Year 4 — Alliums / Root crops / Legumes / Brassicas
Year 5 — As year one

plants such as sunflowers and sweet-corns can provide support for crops such as peas, beans, and squashes; those with leafy ground cover will help to keep the ground cool and moist for others that require those conditions.

There are beneficial influences that are provided by plants underground as well as above ground: some have exudations from the roots that inhibit weeds, for example; some fix nitrogen in the soil, which is extremely beneficial to other plants; root channels made by root hairs and larger roots loosen the soil and provide pathways through the soil for other plants to use, and once the original plant dies it provides nutrients that line these passageways.

Companion planting tends to help most where there is a high density of plants in the soil, thus increasing yield and deterring the presence of weeds through light starvation. It enables one to create a beautiful garden in which the plants live together in harmony and reciprocity—a garden full of color, when marigolds or nasturtiums are interplanted with vegetables, and beds are surrounded by aromatic herbs.

The practice of companion planting can be approached in two ways: either by planting crops in rows (rows of carrots interplanted with rows of onions, for example); or in adjacent beds. It is important to space plants so they all get adequate light and nutrients and can be harvested with ease. There are some minor problems that can occur in using companion plants with different soil preferences. Herbs, for example, tend to prefer a light sandy soil while most vegetables need rich soil with plenty of humus. Cropping times can also differ, meaning that harvesting one crop must be done carefully so as not to damage the companion crop. A good plan needs to be made if this system is to be undertaken methodically and work well.

Plants	Good companions	Poor companions
Apples	chives, garlic, horseradish, nasturtiums, tansy	
Apricots	basil, horseradish, tansy	
Asparagus	anise, basil, parsley, tomatoes	
Beans	anise, brassicas, calendulas, carrots, celeriac, celery, cucurbits, potatoes	fennel, garlic, onions
Brassicas	anise, beetroots, celeriac, celery, dill, garlic, nasturtiums, onions, peas, potatoes, snap beans	radishes, strawberries, tomatoes
Carrot	anise, chives, coriander, garlic, leeks, lettuces, onions, peas, rosemary, sage, tomatoes	dill
Celery	anise, brassicas, beans, leek, tomatoes, dill	
Cucurbits	anise, beans, chives, marjoram, nasturtiums, peas, radishes, sunflowers, sweetcorns	potatoes
Currants	tansy	
Gooseberries	tomatoes	
Leeks	carrots, celeriac, celery, garlic, onions	
Lettuces	anise, calendulas, carrots, cucurbits, radishes, onions	
Onions and garlic	beetroots, lettuces, parsley, strawberries, summer savory	beans, peas
Parsnips	anise, garlic	
Peaches	garlic, nasturtiums, tansy	
Pea	beans, carrots, cucurbits, potatoes, radishes, sweetcorns, turnips	garlic, onions
Potatoes	beans, brassicas, calendulas, horseradish, peas, sweetcorn	
Radishes	anise, chervil, lettuces, mustard, nasturtiums, peas	brassicas
Raspberries	calendulas, tansy	
Spinach	most things, particularly strawberries	
Strawberries	beans, borage, garlic, leeks, lettuces, onions, sage, spinach	brassicas
Tomato	anise, asparagus, basil, calendulas, carrots, chives, dill, garlic, onions, nasturtiums, parsley	brassicas, kohlrabi, potatoes
Turnips	peas	

Vegetables in an ornamental planting

For thousands of years, flowers have been grown alongside principally food crops or medicinal plants—for adorning altars, for their beauty, to combat smells, or, most commonly, because they had medicinal or other virtues themselves. It is possible to change this emphasis in a small garden by incorporating some food plants in a basically ornamental border. This is easily done by leaving a small area in the front of the border which can be prepared for smaller vegetables, such as the ornamental cabbages, lettuces, green beans, basil, and so on. Larger plants (broad beans, scarlet runner beans, climbing green beans, or tomatoes) can be grown on supports further back along with artichokes, which are themselves attractive, tall plants.

There are several advantages to mixed gardening in this way. It not only protects vegetables from pests and diseases, but it is also convenient in very small gardens or where few vegetables are needed.

Container planting

Growing foods and herbs in containers is another and quite feasible way of dealing with a small space—even a patio garden, roof terrace, or balcony. Pots should be placed in a sunny position, near a warm wall or house which will protect the plants from the cold and wind. Herbs are often grown in this way, but many fruits and vegetables can also be grown like this: lettuce, cucumbers, arugula, tomatoes, green beans—for which the pot must be at least 10 in (25 cm) deep—early dwarf peas, strawberries (in strawberry pots), squashes (for which the soil must be loam-based), and peppers. Most herbs are ideal, except for the really large ones such as angelica or borage. A bay looks most attractive in a terracotta pot, and even fruit trees can be grown in containers: dwarf apple trees can be grown outside, while dwarf peaches and even

POT PLANTS

1 Lavender	5 Thyme	10 Dandelions	15 Trailing nasturtiums
2 Sorrel	6 Parsley	11 Arugula	16 Basil
3 Chamomile	7 Chervil	12 Lettuces	17 Coriander
4 Variegated mint	8 Wild strawberries	13 Garlic	18 Chives
	9 Radishes	14 Calendulas	19 Fennel

lemon trees can be kept in a greenhouse in winter and brought outside in the summer. Most bush and cane fruits can be grown in containers if necessary. All like sheltered, sunny places to grow. Use pots filled with organic growing mix. You will need to make sure the plants are watered regularly, and it is a good idea to use water-retaining granules in the mix, as pots dry out quickly in summer.

Pots need to be free-draining, with brick or terracotta pieces or stones placed in the bottom, and are best elevated a little, on bricks or stones, to allow the free circulation of air.

ORGANIC FERTILIZERS

These are good for "quick fixes" for plants; they will not improve the soil in the way that compost or well-rotted manure will, since they provide no bulky material, no humus, and are easily washed out of the soil. They come in the form of liquid manures or powdered concentrates.

Liquid manures

Comfrey (Symphytum officinale) or Russian comfrey (*Symphytum uplandicum*), which is richer in minerals, makes an excellent weekly feed for potatoes and tomatoes. Roughly fill a plastic or fiberglass water barrel with cut comfrey (you can get 3–4 cuttings a year from your plants), and fill the barrel with water. A barrel with a faucet at the side is best. After 2 to 4 weeks you will get a very black, slimy (and smelly) mess. This liquid can be strained off to provide a very concentrated feed, rich in potassium, which should be diluted 20 parts water to 1 part comfrey liquid.

Nettle liquid is so nutritious (only low in phosphorus) that it is nearly a complete fertilizer. It is made by the same process as for comfrey, using about 2 lb nettles to 2 gallons water (1 kg nettles to 10 liters water). Nettles collected in spring are best. Strain off the liquid after 2 weeks and dilute 10:1. If it is used when transplanting, to water in, diluted nettle liquid will boost growth and further applications will encourage abundant fruiting.

Seaweed liquid is good for tomatoes and also where plants are showing signs of any sort of deficiency. It contains all the trace elements and can be sprayed onto the leaves or fed directly into the soil.

Powdered concentrates

Rock phosphate should be used in preference to "superphosphates." This provides slow-release phosphorus where it is lacking in the soil. It is unlikely to be needed in most gardens—only those in acid areas of high rainfall, on deep peat, on very heavy clay soil, or where soil has been damaged by the overuse of chemical fertilizers. It is generally applied in autumn—8 oz per sq yard (240 g per sq meter). It is not, however, ideal as it is expensive and comes from finite sources. Bonemeal can be used as a substitute.

Bonemeal provides slow-release phosphorus, plus calcium and various trace elements, and makes a good top-dressing for strawberries. Do not use anywhere near calcium-hating plants.

Dried blood provides long-lasting slow-release nitrogen to boost plant growth. It is useful in the spring and summer and

A TRADITIONAL VEGETABLE GARDEN

1 Lavender	15 Sorrel	27 Vegetable bed for rotation 1 (brassicas)	37 Rhubarb
2 Lawn	16 Lemon balm		38 Fennel
3 Bay tree	17 Mint	28 Raspberries	39 Rosemary
4 Runner beans	18 Sorrel	29 Black and red currants	40 Pumpkins and squashes
5 Bush Basil	19 Caraway		
6 Purple Basil	20 Knotted Marjoram	30 Gooseberries	41 Asparagus
7 Chervil	21 Thyme	31 Blackberries	42 Strawberries
8 Chives	22 Garlic	32 Cucumbers and tomatoes	43 Spring onions
9 Dill	23 Sage		44 Pears
10 Fennel	24 Chamomile	33 Compost heaps	45 Cloches
11 Coriander	25 Vegetable bed for rotation 1 (roots)	34 Apples	46 Potatoes
12 Parsley		35 Horseradish	47 Lettuces
13 Oregano	26 Vegetable bed for rotation 1 (legumes)	36 Artichokes	48 Arugula
14 Borage			49 Chicory

can either be added to water and used as a liquid feed or mixed into the topsoil at 1 oz per sq yard (30 g per sq meter). *Greensand,* also known as glauconite, is an organic source of potassium, which is released slowly into the soil. Broadcast it over the soil and lightly rake it in.

PESTS AND DISEASES

An organic garden provides a habitat for both pests and their predators. It is important to remember that if the soil is healthy the plants will usually have fairly good resistance to pests and diseases. When buying seed, always choose resistant varieties, and when buying plants, make sure there is no discoloration of the leaves or sign of disease.

Attacking pests with chemicals kills not only the pests but also a good many of their predators. It destroys the delicate ecology of the environment, restricting the food supply of natural predators. In the long run, chemical pesticides only exacerbate the pest problem; they also drench our food and the soil in poisons.

There are some general preventive measures that can be taken against pests that will help keep a good natural ecological environment in the vegetable garden:
1 Keep the soil healthy with good compost or manure, and always remove and discard any diseased plant material.
2 Practice crop rotation (*see pages 132–3*), which prevents soil-borne pests, spores, and fungi from getting a real grip on the crops. If you do not have room for a formal crop rotation, make it a general principle not to plant the same crop in the same place two years running.
3 Practice companion planting (*see pages 133–4*), using other plants to deter pests.
4 Avoid growing large areas of any one thing (monoculture), which promotes a tendency to insect infestations.
5 Grow plants that attract natural predators such as ladybugs and hoverflies

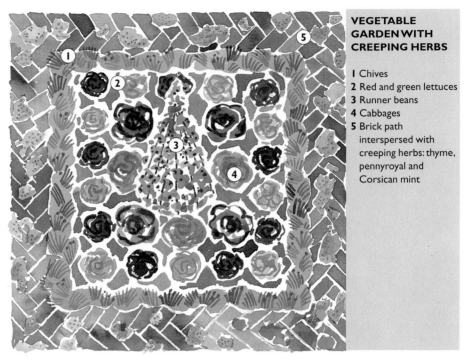

VEGETABLE GARDEN WITH CREEPING HERBS

1 Chives
2 Red and green lettuces
3 Runner beans
4 Cabbages
5 Brick path interspersed with creeping herbs: thyme, pennyroyal and Corsican mint

which feed on aphids. Nasturtiums, for example, attract hoverflies; sunflowers attract bees, lacewings, and predatory wasps; mints attract hoverflies and predatory wasps; and dill attracts wasps and hoverflies.
6 Avoid treading on ground beetles, which like to live under rocks. They will usefully eat eelworms, cutworms (a species of caterpillar that attacks vegetables), and leatherjackets (the larvae of daddy longlegs that feed on plant roots).
7 If you have a pond and a wild area of the garden, this will encourage frogs and toads, which will, in turn, eat snails, sowbugs, and wireworms.
8 Birds can be a help or a hindrance, but chickadees and house wrens will happily pick off aphid eggs, and others will eat slugs and snails. So, encourage birds to the garden by providing food and water for them in winter, but protect young crops with wire netting.
9 Pick up caterpillars by hand and throw them to the birds. Wear gloves as some caterpillars may give you a rash.
10 Sink a jar or saucer of beer in the ground to attract slugs and snails, or go

out at night with a flashlight and capture them by hand. Put the slugs and snails into strongly salted or boiling water to kill them (*and see use as garden spray in "Natural pesticides and fungicides," page 138*). If you are squeamish about killing them, dump them in a field well away from the garden.
11 Learn to tolerate a few bugs and holes in your lettuce, as long as they leave enough for you. Just wash your vegetables thoroughly or soak them briefly in salted water.
12 Carrot rust fly (whose larvae also eat the roots of parsley, parsnips, and celery) cannot fly very high. A very close mesh, at least 30 in (75 cm) high around the vegetable patch will keep them away, and so will floating row-cover fabric laid over the plants and anchored at the sides.
13 Old carpet or cardboard placed on the ground around the stalks of young brassicas will prevent cabbage root flies from laying eggs. Cut 4–6 in (10–15 cm) squares and make a slit to the center.
14 Put plastic collars around the base stems of zucchinis to help protect them from cutworms and slugs. You can make

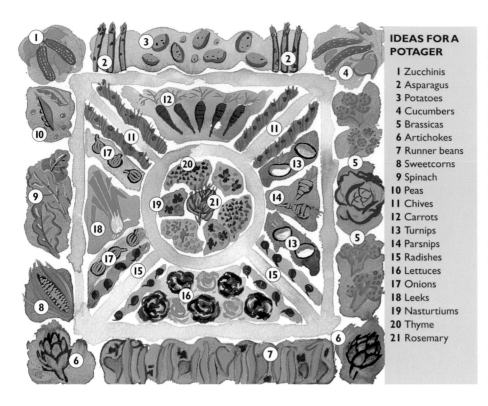

collars from cans with the tops and bottoms taken out, or plastic drink bottles or yogurt containers, topped and tailed.

15 English holly leaves scattered between rows of crops will deter mice. So will summer savory, mint, and narcissi planted near beans.

16 Sow vegetable seeds at a time most likely to avoid pests: for example, carrots in early summer rather than late spring, to avoid carrot rust fly.

17 Place tomato leaves between rows of vegetables to drive away flea beetle.

18 Ensure planting conditions are suitable (light or shade, soil type, etc.) to produce hardy plants.

19 Avoid planting brassicas in summer if clubroot is a problem. Crop rotation will also help. If vegetables are affected, use a heavy dressing of lime on the plot in autumn before sowing the following year. Beware of excess lime in the soil, which can predispose to canker in beets and affect cauliflowers, turnips, and tomatoes.

20 If there is a really persistent problem, simply avoid growing that particular crop for a year or two.

NATURAL PESTICIDES AND FUNGICIDES

Garlic or onion spray is effective against blackfly, carrot rust fly, wireworms, pea and bean weevil, slugs, and onion thrips. Chop about 4 oz (115 g) of garlic or onion and let it macerate in a little oil overnight. Then press and strain, reserving the oil. Add the oil to 1 pint (570 ml) of water in which you have dissolved 1/3 oz (10 g) of insecticidal soap and then dilute 1 fl oz (30 ml) in 1 pint (570 ml) of water and use as a spray.

Tansy spray is useful against aphids, cabbage worms, Colorado beetles, Japanese beetles, and squash bugs. Make up a double-strength infusion and use it neat.

Slug and snail water will deter just about everything. After several days, filter the water in which the slugs and snails were killed and use the liquid as a general garden spray.

Insecticidal soap will dissolve the waxy shell of aphids and kill them, but does no harm to their predators like ladybugs and hoverflies. It also kills red spider mite and whitefly. Insecticidal soap can be

bought from your local hardware store or garden center. Dissolve 1 oz (25 g) in 6 pints (3.4 liters) of rainwater, or follow container directions, and use it as a spray. *Rotenone-pyrethrum* will kill greenfly, blackfly, thrips, aphids, Japanese beetles, cucumber beetles, and flea beetles. It can be applied as a liquid spray (dilute as specified on the bottle) or as a dusting powder. **Caution:** This biodegradable plant-derived powder is a poison and must, therefore, be used carefully, wearing protective clothing. It is toxic to fish, as well as to bees, although it is not harmful to plants.

Bordeaux mixture is a copper-based mixture, useful for cases of potato blight and leaf spot. Purchase it as a powder, mix it according to the instructions given, and use it as a spray. You should, however, remove and discard the infected parts of affected plants. **Caution:** Be careful when using Bordeaux mixture because it is poisonous to humans.

Seaweed solution helps to control aphids, potato scab, leaf curl virus, damping off in seedlings and tomatoes, and potato eelworms. Use a weak solution at two-week intervals.

Rhubarb or elder spray is used for killing greenfly and other aphids as well as small caterpillars. Boil 1 lb (450 g) of rhubarb or elder leaves in 5 pints (2.8 liters) of water and simmer for 30 minutes. Let cool. Strain and use as a spray.

PLANTING FRUIT TREES AND BUSHES

Fruit trees and bushes in the garden can be decorative, utilitarian, and productive. You can plant fruit bushes such as blackcurrants to make hedges and espalier or fan-shaped fruit trees can decorate walls, separate one area of the garden from another, or hide an unsightly compost heap or garden shed.

When buying fruit trees or bushes,

make sure you choose healthy specimens from a reputable nursery or reputable garden center, and choose varieties that are suitable for your purposes, the space available, and your soil and general conditions.

Fruit trees and bushes should ideally be planted out during their dormant season in late autumn. They then have time to become established before putting their energy into new spring growth. If that is not possible, plant them no later than early spring. Never plant in frosted ground and try to avoid planting them in a frost pocket.

Before planting, the ground should be well prepared and cleared of all perennial weeds (*see pages 125–7*). Dig in good organic matter and remember that the ground near a brick or masonry wall is usually drier than in other parts of the garden, as the wall absorbs moisture, so mulch this area extra well. You will need to put up support wires for cordons, fans, or espaliers (*see below*). For these, use vine eyes available from garden centers

and mail-order sources to attach the wire to the wall, and make sure it is tightened sufficiently. You may also need to erect strong, sturdy posts and wire for growing raspberries.

If you decide to grow fruit trees or fruit bushes as fans, espaliers, or cordons against a wall, certain varieties grow best in different situations:

Against an east-facing wall: pears, cherries, redcurrant, and gooseberries.
Against a west-facing wall: plums, cherries, apples, and pears.
Against a south-facing wall: blackberries and, in milder climates, peaches and apricots.
Against a north-facing wall: redcurrants, and gooseberries.

FRUIT TREES

These can be grown as standards (traditional orchard trees), half standards, dwarf pyramids (free-standing, and somewhere between a bush and a cordon), or cordons. New varieties of apples with single stems that fruit on spurs are

particularly suitable for container growing and small spaces, but they must have plenty of light.

All fruit trees are now grafted onto the rootstocks of other varieties, and it is these that determine the eventual size of the trees. If intending the tree for a small space, always check that dwarfing stock has been used, otherwise you may well end up with an enormous tree. Many can also be trained to be used as a screen or to go against a wall. Cordons, fans, or espaliers can be used here.

Cordons take up the least space and consist of a single stem with short spurs. They are either trained on wires against a wall and usually grown obliquely, or are planted free-standing, supported with posts, which should be buried 2 ft (60 cm) deep in the ground and spaced about 9 ft (2.7 m) apart. The support wires should be fixed 18–24 in (45–60 cm) apart.

Espaliers can be grown against a wall, or as a screen on a good support structure of free-standing posts and wires, as for a cordon (*see above*). They are trained to

(continued on page 142)

Stepover

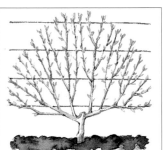

Fan trained

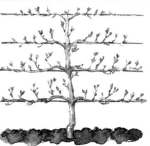

Espalier

Standard dwarf

Cordon

Two types of tree supports: wires and vine eyes against a wall (right), and wires and posts in an open garden (far right).

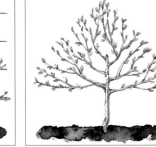

Planting a bare-rooted fruit tree or bush

1 *Soak the roots well for several hours before planting. Check for damage and cut away damaged or any that are much longer than the others with a clean cut that will point downwards when the tree or bush is planted.*

2 *Dig a hole deep and wide enough to accommodate the roots without cramping them. Hold the plant in the hole to check if it is large enough; if not, make it bigger. Trees should be planted at the same level as they were in the nursery, so look for the soil mark on the stem. Lay a cane across the hole to ensure planting to the correct depth.*

3 *Break up the soil at the bottom of the hole with a fork to allow the circulation of air and water. Fork in some compost and cover it with a layer of soil.*

4 *If the tree or bush is to be staked (see left), hammer the stake firmly into the ground to the windward side of the plant before planting. If it is put in afterward, the roots may be damaged. If a newly planted tree or bush moves in the wind, the tiny hairs on the roots will break and the plant will be damaged. Hold the tree in the hole, spread out the roots, and gently shovel fine soil back over them.*

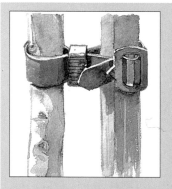

Your support must be able to take the weight of a fully developed plant, which may be considerable, particularly when it is subjected to rain, snow, or wind. Attach the tree or bush to its stake. Use an adjustable tie (above left)

available from garden centers, or protect the stem with burlap and use garden twine tied firmly to the stake and then tied loosely round the tree (above right). Do not use plastic-covered wire as it may cut into the plant.

5 *Fill the hole with soil and tread it down gently to avoid damaging the roots. Level the soil to prevent puddles forming. Water well. Check over the next few days to see if the soil has settled. If it has, then top it up and firm it down again.*

Planting a container-grown tree or bush

1 *Dig a hole wide and deep enough to accommodate the root ball of the tree or bush.*

2 *Loosen the soil at the sides and bottom with a fork. Incorporate some compost and cover it with a layer of soil.*

3 *If the tree or bush needs to be staked, hammer the stake very firmly into the ground to the windward side of the plant before planting.*

4 *Ease the tree or bush gently out of the pot and gently separate any tangled roots, being careful not to break up the root ball.*

5 *Hold the tree or bush next to the stake in the hole and distribute soil around the roots with your fingers. Lay a cane across the hole to ensure that you are planting to the correct depth by referring to the original soil mark.*

6 *Fill the hole with soil and tread it down very gently to avoid damaging the roots. Level the soil so that no puddles can form. Water well. Check over the next few days to see if the soil has settled. If it has, then top it up and firm it down again.*

Fruit cages and nets

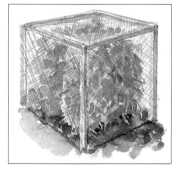

You may need to protect fruit trees and bushes from frost, deer, mice, and birds. Cover them in winter with burlap, row-cover fabric, or fine mesh netting to protect them from the cold. To protect against birds, use a fine mesh netting, or things that flutter in the wind. A fruit cage (far left) made of wire mesh on a wood frame is best for soft fruit bushes such as black currants. Wrap the base of tree trunks with plastic tree tape to keep mice away from the tender bark. Sturdy fencing is required to keep deer away.

1 *Choose a firm, ripe stem and take cuttings of at least 9in (23cm). Cut straight across the stem beneath a node.*

2 *Cut away all the leaves and trim to approximately 6in (15cm), cutting at an angle directly above a node.*

3 *Dig a narrow trench, which should be 1–2in (3–5cm) shallower than the length of your cuttings.*

4 *Sprinkle a little sand into the bottom of the trench to improve drainage.*

5 *Place the cuttings in the prepared trench at intervals of about 4–6in (10–15cm). Remember that the straight-cut end is the bottom of the cutting and the angled end is the top.*

6 *Fill in the trench and firm the soil around the base of the cuttings. Water well and label the cuttings clearly.*

(continued from page 139)

rise in series of horizontal tiers (usually four) from a central stem.

Fans radiate out against a wall or screen, and new growth is trained to fill in the spaces. Support wires should be spaced about 6 in (15 cm) apart.

Stepovers, which are a form of espalier that is suitable for apples and pears and are grown as low as 1 ft (30 cm), make a lovely edging for paths. If you buy trees that have already begun to be trained (espaliers with two tiers, for example, or fan-shaped with four ribs) this will save you a lot of trouble.

Since most fruit trees require a pollinating partner, you will need to buy two trees that flower simultaneously. When buying plum or pear trees, you will need to check that they are compatible.

When planting soft fruit, it is best to try to avoid frost pockets since the blossoms are fragile and easily damaged. Any damage means that you will not get fruit later in the season from those flowers. When buying plants, find varieties that are frost-resistant, and check that they are free of viruses. When growing blackberries and raspberries, you will need posts and wires to support them (*see page 139*)—three wires strung in parallel lines between posts 4 ft (1.2 m) apart will support raspberries, while five wires are better for blackberries. Blueberries, black currants, red currants, and gooseberries can be grown free-standing. Red currants and gooseberries can also be grown as cordons against a wall or fence, or supported by posts and wires.

PRUNING FRUIT TREES

Before pruning fruit trees of any type, it is essential to check the requirements of each variety and know exactly when it fruits before you pick up your pruning shears. A good general principle of pruning is: "to stimulate new growth, cut out old wood." Always use sharp pruners and cut cleanly when pruning. Although some pruning should take place in winter, it must never be attempted during frosty weather.

Correct pruning will control the size of the tree, improve its overall shape, and encourage new growth. However, you must check the flowering times of the different varieties and know the fruiting habits of each species: if, for example, the tree fruits on new growth, you can cut out the previous year's growth; if, however, the fruit grows on the previous year's growth, you must be careful not to remove growth on which fruiting will take place. Free-standing trees seldom need to be pruned vigorously; it is enough to remove diseased stems and thin out stems that are severely overcrowded or crossing.

The correct and rigorous pruning of trained fruit trees is, however, essential. With these, any outward-growing shoots need to be cut off, but other training will depend on the style chosen. Pruning will strengthen the root system, allow more air and light to circulate (reducing the incidence of disease) and rain to penetrate, and improve both the size and the quality of the fruit. When pruning, thin out the weaker branches, removing lateral stems that may damage the tree by rubbing against other stems.

Winter pruning is undertaken to encourage new growth in the following spring. It consists of cutting out dead wood and thinning overcrowded stems. Leaders are best shortened to a strong, upward-facing bud, which will then spring to vigorous life as soon as the warmer weather commences. Winter pruning is usually undertaken in late autumn to late winter, but in colder regions it is best to wait until early spring.

Summer pruning is used especially for cordons, espaliers, dwarf pyramids, and fans. It aims to reduce leaf growth and so stimulate the tree into making fruit buds. It should be done in mid to late summer. The leaders should be shortened by a quarter of their new growth and the laterals by a third of their new growth. In the following winter, take off all the new growth that has occurred since the summer pruning and remove one further bud of the old growth. Aim to improve the shape of the tree overall, and if lopping a branch to remove it completely, make sure that the cut is completely even, leaving just a small collar to aid healing. At the same time, tie in loose branches.

Pruning soft fruit bushes
This is done in late autumn when the leaves have dropped and it is easy to see any dead wood. Either remove these dead stems entirely or cut them down to a new, strong sideshoot. Cut healthy stems down to a third or fourth outward-growing bud. Bushes should be cut into what is known as the "goblet" shape, on the principle that the center should be open to allow light to penetrate.

Red currants live up to 25 years and need no drastic pruning. The fruit grows on spurs made on old wood. When pruning red currant or black currant, set

Tip fruit bud
A large, rounded bud that produces first blossom and then fruit at the tip of a branch.

Sub-lateral
A side-shoot that grows from a lateral.

Leader
The leading shoot of a main branch.

Water shoot
A vigorous shoot that is unfruitful and should be removed.

Growth bud
Small, flat bud that produces leaves or a new shoot.

This year's growth
Shoot development that has been made during the current growing season.

Spur fruit bud
Large rounded bud on a spur that produces first blossom and then fruit.

Spur
A short lateral branch that bears fruit buds.

Lateral
A side-shoot that grows from a leader.

PRUNING CUTS

✔   ✗   ✗   ✗

healthy pruned stems aside. Make a trench in your vegetable garden and plant a row of cuttings—about 8 in (20 cm) long—with only 2 in (5 cm) of the cutting showing above ground. Firm the earth round the cuttings, water them, and leave them until the following spring. With any luck you should have a row of fresh, new plants ready for planting out.

PROPAGATION

BUYING AND SAVING SEED

Most vegetables are raised from seed that is either bought or saved from the previous year's crop. It is always fun in winter to browse through the seed catalogs, decide what you are going to grow, make crop rotation and companion planting plans, and place your order. When choosing seeds for your garden, you need to consider the geographic location of the area, the location of the plot, and the type of soil you have to work with. Choose appropriate varieties that are disease-resistant and reliable varieties that you know will grow in your garden; but you may also want to experiment with others.

When saving your own seed, make sure you collect seeds from healthy plants. Peas and beans are easy to collect. Leave some pods to ripen on the plants, pick and shell them, and put the dry seeds in clearly labeled envelopes for the following season. Onion and leek seeds should be dried in an airy place and then stored in labeled envelopes. Do not attempt to save seeds from hybrid varieties, since they will not grow true and you will get a "throwback" to one or more of the parent varieties. Concentrate instead on the "heirloom" varieties for successful seed-saving.

SOWING SEED

Instructions printed on all seed packets give basic directions about the time and depth of sowing. The time of sowing, however, will be subject to some variation, depending on where you live and the weather conditions at the time. It is better to wait a week or two for better conditions than to sow when instructed if it is too wet, too dry, or too cold. Seeds sown "late" will catch up and do better than those sown under poor conditions.

Basically, most seeds need moisture and warmth in order to germinate, but not a wealth of nutrients at this stage.

However, some vegetables, such as lettuce and spinach, prefer cold soil if they are to germinate successfully.

Most vegetables can be successfully sown out of doors, either in seed beds (like the brassicas, which will be transplanted into a plot in autumn once the summer crops are over), or in the place in which they are to grow. However, seeds sown indoors will give you an early start, especially with the more tender plants, and then they can be planted out after the last frost—usually at the end of spring—but this will depend on the year and your location. Suitable plants for optional indoor sowing include snap and scarlet runner beans and summer squash. Some plants, such as cucumbers, tomatoes, and peppers, need the higher temperatures indoors or in the greenhouse in order to germinate.

Sowing seeds in seed trays indoors is fairly labor-intensive since the seedlings will need to be pricked out, potted up, and then hardened off (*see pages 145 and 146*). There are various types of growing mediums you can use. Most proprietary seed and potting mixes contain loam, sand, and peat, as well as chemicals; organic ones are available, and it is worth looking for them. You can also make your

Sowing smaller seeds in seed trays or other containers

1 *Fill a seed tray with well-watered mix and firm gently with a board or the base of another tray.*

2 *Sprinkle seeds over the surface, cover with a shallow layer of mix, and water the seeds gently.*

3 *Label the tray with the name and date of sowing. Cover the tray with glass, polyethylene, or newspaper to*

keep in the moisture. Shade the tray if in direct sunlight. Some seeds germinate best in the dark (check the directions on the seed packet). Check every day that the glass or polyethylene has not collected too much condensation; seeds may go moldy if they become too damp. As soon as the seeds have started to germinate, remove the covering in order to give the seedlings plenty of air and light.

Sowing larger seeds in pots

1 *Fill a pot with well-watered starter mix. Press the surface down gently, using the base of another pot.*

2 *Push 4–5 seeds into the mix ¹/2–1 in (1.5–2.5 cm) deep.*

3 *Gently sprinkle and firm mix over the seeds.*

4 *Water gently, using a watering can preferably fitted with a rose. Label with the name and date of sowing.*

1 *Fill a pot, compartmented pack, or other container with well-watered potting mix and make small, round holes with your finger or a stick.*

2 *Ease the seedlings up gently, using a small knife or spatula, and gently separate them. Take care not to damage their roots.*

3 *Hold the seedling by a leaf (avoid touching the stem if possible) and place it in a hole in the new pot so that the roots fall easily into the hole.*

4 *Gently firm the soil around the seedling and water, using a watering can preferably fitted with a rose.*

own homemade seed and potting mix (*see below*), but never use garden soil because of the risk from soil-borne diseases and pests.

Sowing seeds indoors

You can sow your seeds either in bought or homemade seed-starting mix, in seed trays, seed modules, pots, or plastic tubs from the kitchen (with holes made in the bottom for drainage), following the instructions found on all seed packets.

Smaller seeds are usually sown in seed trays; larger seeds, such as beans and squashes, should be sown 2–3 to a pot. Soil cubes, which are made with a gadget like a cookie press, have become popular in recent years because they do away with the need for pots altogether.

Homemade seed-starting mix

If you want to make up your own seed-starting mix, it is recommended to use fine leafmold 2–3 years old, either on its

own or with some sifted loam from turf stacked upside down and left for a year. Alternatively, you could use 4 parts comfrey leafmold (*see page 130*) to 1 part sharp sand, or 2 parts peat moss and 1 part horticultural sand soaked in liquid seaweed (*see pages 126 and 136*).

Homemade potting mix

Seedlings come up very thickly and must be pricked out and potted up in larger containers with more nutrient-rich potting

Planting out or transplanting young plants

1 *Use a line or string tied to two sticks, a hoe handle, or plank to form a straight line, and use a dibble or trowel to make a row of holes in the soil.*

2 *Hold each seedling carefully by a leaf, lower it into a hole and gently firm the soil around each one.*

3 *Tender young plants can be protected from wind, slugs and snails, and from drying out for the first few days by covering them with cloches. These can easily be made by cutting the bottom off plastic soda bottles and placing the top part over the plant; remove the screw top to allow for ventilation. Another way to protect young seedlings from insect and bird attacks is to use an ultra-light horticultural fabric called floating row cover. Gossamer-thin, it comes in rolls of various widths and is so light that plants push it up as they grow. Light and water can both penetrate, but even the tiniest insect is kept away from the plants.*

mix before they become leggy and weak from insufficient light, air, and space.

Potting mix, like seed-starting mix, can also be bought or homemade. Here are two recipes you can try yourself:
1 4 parts loam, 2 parts peat (or peat alternative, such as 2–3-year-old leafmold), 8 oz (225 g) seaweed meal, and 4 oz (115 g) bonemeal.

2 In a clean garbage can, mix 3 parts peat (or peat alternative), 1 part sharp sand, and 1 part compost. Add 3 oz (85 g) calcified seaweed to every 10 gallon (45 liters) of mixture.

Hardening off
Young plants ready for planting out should first be hardened off to protect them from the shock of going into direct sun, cold soil, cold night air, or poor weather conditions after the protected environment indoors. It can be done in several ways. The pots can be put under cloches or in a cold frame outside, or put out in the open during the day and brought in at night.

Planting out or transplanting young plants
The best time for planting out or transplanting young plants is the evening. If you plant out in the middle of a hot summer day the plants will wilt and not get off to a good start. Make sure the soil is moist and well prepared and that the plants are already watered. A dibber is useful for making holes for new plants. Make sure that the hole you make is deep enough not to cramp their roots. It is important that tomatoes, squashes, and sweetcorns are planted with a trowel because the rootball should be left intact.

You can, of course, buy seedlings from a garden center and plant them out. Make sure you buy plants with healthy dark green leaves, checking for signs of disease and rejecting any that are not

PLANTED PATIO

1 Lettuces
2 Zucchinis
3 Sage
4 Peas
5 Carrots
6 Rosemary
7 Thyme
8 Onions
9 Beets
10 Dwarf beans

strong. Do not buy any brassicas with swelling roots (this is an indication of clubroot disease).

Each small plant should be handled carefully by a leaf so as not to damage the stem or disturb the root system. Always make sure that seedlings are well watered before planting in and that the ground is well prepared.

When planted in rows, seedlings can be staggered, allowing you to fit more of them in the space. This also gives each one more leaf space without it encroaching on its neighbors. It also helps to suppress weeds by depriving them of light.

Sowing seed out-of-doors in seed beds and in situ

Ideally, having dug garden compost and/or well-rotted manure into the soil the previous autumn, fork the soil over lightly and break up any large lumps of earth. Tread the soil down to break up the lumps further, then rake it over very gently with a backward and forward motion rather than dragging the rake over the soil, until there is a fine tilth. Repeat these stages if necessary until a fine tilth is achieved.

- Small seeds (carrots, lettuce, parsley) need a groove ½ in (1.5 cm) deep.
- Slightly bigger seeds (spinach, beets) need a groove 1 in (2.5 cm) deep.
- Pea and bean seeds need individual holes 3 in (7.5 cm) deep (this can be done with a dibber).
- Brassicas should be sown in seed beds on their own; later in the year the young plants can be planted out to replace earlier crops that are over in the vegetable plot.

Always remember to label rows of seeds and to protect them from birds by using wire netting, pieces of string tied to sticks and zig-zagging 6–9 in (15–23 cm) off the ground, bits of kitchen foil or brightly colored plastic fluttering in the breeze, Christmas decorations, scarecrows, or anything else you can conjure up as a bird scarer.

Thinning out

Seeds sown in rows usually come up far too close together and the seedlings will then need to be thinned out in order to allow room for the young plants to develop. This will probably need to be done more than once. The first thinnings can be discarded as the seedlings will be too small to transplant or to eat.

Never leave thinnings lying on the soil as they attract pests; this is especially true of onions and carrots, which attract onion thrips and carrot rust flies. Thin the seedlings carefully, and gently press the soil back over the remaining plants.

The second thinning often produces miniature vegetables that are very sweet and delicious to eat; some of these thinnings may even be worth transplanting to another site or giving to a friend.

Young plants can be harvested as needed from the rows, which will again leave increased space for the remaining crop to develop.

Sowing

1 Use a line or piece of string tied to two sticks to mark out where you are to sow your seeds. Alternatively, use a hoe handle or a plank to provide a straight line.

2 Make a groove along this line, either with a stick or with the corner edge of a hoe, to a depth suitable for the particular seeds.

3 Gently water the groove using a watering can, preferably with a rose, then trickle fine seeds through your fingers to distribute them evenly along the groove. It helps if you first mix very small seeds with fine, dry sand. Cover with soil.

4 Draw the rake carefully across the groove and gently tread the soil down, aiming to get the earth as flat as possible to avoid any dips where water can collect.

storing produce

Preserving and storing your own homegrown produce is worthwhile and satisfying on many different levels. Apart from following in a tradition that stretches back for thousands of years to a time when preserving food was essential for survival, the range of different storing and preserving techniques gives you the opportunity to enjoy your produce at any time of the year, in or out of season, and to create tastes and flavors that, once again, make food a positive pleasure.

Preserving food

The traditional kitchen gardener has always aimed to grow produce to provide a tasty and interesting selection of fresh, organic food for as much of the year as possible. By storing and preserving the harvest, the garden's bounty can be enjoyed year-round.

The availability of freezers has led to a decline in the use of such traditional preserving methods as canning, pickling, and drying, but there is satisfaction in having shelves loaded with gleaming bottled fruits, preserves, and pickles, and storage areas filled with fruits and vegetables. Dried fruit is a tasty and nutritious snack, too. Stored in these ways, your produce is safe against power outages that would affect a freezer.

Produce will not improve from being stored, canned, or dried, so start with the pick of your crop. Avoid bruised or damaged items, or anything that shows even the slightest sign of mold or insect damage. One bad piece can spoil the whole batch.

CANNING

When canning, bear in mind that dangerous molds and bacteria can grow and thrive in improperly canned produce. It is really easy to do it the right way, so why take chances?

EQUIPMENT

In addition to regular kitchen equipment, the tools for safe canning are few and mostly inexpensive. You may already have some in your kitchen. You will need:

- Boiling-water canner with a wire rack and tight-fitting lid—useful for other things as well as canning.
- Jar lifter—special tongs for handling jars safely.
- Canning funnel—makes filling easier, and helps to keep jar tops clean.
- Metal tongs—for removing things from boiling water.
- Ladle—for filling jars.
- Narrow plastic spatula—to remove air bubbles.
- Steam-pressure canner—a more expensive item, but essential if you want to can low-acid foods.

CONTAINERS

Canning jars come in several styles and sizes. "Quilted" jars make attractive gifts, and straight-sided, wide-mouth jars can be used for canning or freezing; jars with shoulders may burst if they are frozen. Pint and quart sizes are good for most home-canning purposes.

Do not be tempted to use old mayonnaise or jelly jars, made for one-time use, or old-fashioned canning jars with spring closures. To ensure a safe seal, buy new, sturdy canning jars (often called Mason

jars) with separate lids and screw bands.

The jars themselves (if they are undamaged) and the screw bands (if they are free of corrosion) may be reused. But do not reuse canning lids—the rubbery compound that provides a safe seal may not work the second time. Jars, lids, and caps need to be very clean or the produce is likely to spoil. Wash jars and lids in warm soapy water and rinse thoroughly. Do not use any kind of abrasive cleaner. Place the jars in the canner, half fill it with water, and bring it to a boil for 15 minutes. Reduce the heat to a slow simmer and leave the jars in the hot water until you are ready to use them. Place the lids in a pan of hot (but not boiling) water until you are ready for them. Dry the screw bands and set them aside.

FILLING JARS

It is best to deal with one jar at a time— fill it and immediately fit its lid before moving onto the next one. This reduces the possibility of bacteria infecting the produce. Using a jar lifter, remove the jars from the hot water, empty the water out (let it drain for a couple of seconds) and place the hot jar on a clean towel on your countertop. Immediately fill the jar with the prepared hot produce.

Using a canning funnel helps to keep the top of the jar clean as you fill it. Even so, you should wipe the top of the jar with a clean paper towel before placing the lid in position.

To ensure a proper seal, it is important to leave the correct space between the top of the food (or its liquid) and the lid of the jar. This is called the headspace. A general rule of thumb is to leave a 1 in

(2.5 cm) headspace for low-acid foods and vegetables; a ½ in (1.25 cm) headspace for high-acid foods, tomatoes, and fruit; and a ¼ in (65 mm) headspace for jams, jellies, and pickles.

After filling a jar, use a narrow plastic spatula to remove any air bubbles that may have become trapped in the produce. Do not use a metal knife to do this, since its hard edges could scratch or otherwise damage the hot jar.

Carefully remove a lid from the hot water and center it on the jar. Centering the lid ensures that the rubber composition seal will work properly. Then fit the screw band.

PROCESSING

Different foods require different processing times. A boiling-water canner reaches 212° F (100° C), which is fine for acid foods. A steam-pressure canner reaches 240° F (115° C), which is the higher temperature needed for low-acid foods. These two methods are not interchangeable, so follow the recipe instructions.

Boiling-water canner

Fill the canner with enough warm water to cover the jars. After lowering the filled and capped jars onto the wire rack in the water, put the lid on and turn up the heat to high. Bring the water to a full boil. Adjust the heat to keep the water at a gentle rolling boil. Start timing the processing period called for by the recipe.

When the time is up, turn off the heat and remove the canner lid. Be careful of the scalding hot steam. Using the jar lifter, gently remove the jars and place them on the towel on your countertop. Allow about 2 in (5 cm) of space between the jars. Leave them there to cool, undisturbed, for 12 to 24 hours. Avoid drafts

that might crack the hot jars. Do not tighten the screw bands.

Steam-pressure canner

Put 2–3 in (5–7.5 cm) of water in the bottom of the canner and heat it to a simmer. Using a jar lifter, place the filled, capped jars on the wire rack. When the rack is full, lock the canner lid in place. Following the manufacturer's instructions, raise the heat to medium high and allow the steam to vent for 10 minutes. Close the petcock or replace the weight and allow the canner to come up to the correct pressure.

Once the desired pressure has been reached (see recipes), start timing the processing period. When processing is finished, turn off the heat and let the canner cool. *Do not open the petcock or remove the weight until the canner has depressurized to zero.* Then wait for

Preserving acid fruit

1 Place jars in canner, with enough warm water to cover it and boil for 15 minutes.

2 After boiling, remove one of the jars, drain it, and place upright on clean paper towels.

5 Wipe the top of the jar clean and place and center the hot jar lid. Fit the screw band and tighten it until you feel resistance. Don't use force. Now fill and put lids on the remaining jars, one by one.

3 Using a ladle and a wide-mouth funnel, fill the hot jar with the prepared produce.

6 Place the jars into the wire rack and when it is full, lower it into the water and process for the correct time.

4 With a plastic spatula, gently pack the fruit down to dislodge any air bubbles.

2 minutes before removing the lid, and protect yourself against the escaping steam. Leave the jars in the canner to cool for 10 minutes. Using the jar lifter, gently remove the jars and place them on the towel on your countertop, allowing about 2 in (5 cm) of space between the jars. Leave them there to cool, undisturbed, for 12 to 24 hours. Avoid drafts that might crack the hot jars. Do not tighten the screw bands.

When the jars have cooled, check the lids for a good seal. The lids should be concave. If you press the center of the lid and it moves up and down *the jar is not sealed*. If this happens you can either store the jar in the refrigerator and use the contents soon (within, say, a week) or you can reheat the produce, repack it in clean, sterilized jars, and reprocess it. If a jar did not seal, it is probably because the headspace was incorrect or the tops of the jars were not clean—some food or moisture interfered with the seal.

STORING CANNED PRODUCE
Label the jars with the contents and date and store canned produce away from direct sunlight. An old-fashioned country larder is ideal for this—cool, dark, and dry—but a cupboard may provide similar conditions. Canned produce, pickles, jams, and preserves will usually last, if prepared and stored properly, for a year or more. Once opened, they must be kept with a lid on in the refrigerator.

Caution: If you open a jar of canned produce and the contents have a bad color or odor, or if the contents spurt out under pressure, throw them away immediately. Then thoroughly wash the jar, your hands, and anything else that has come in contact with the spoiled produce. Botulism is dangerous!

RECIPES
Foods for canning are divided into two groups: acid and low-acid. Acid foods include most fruit (including tomatoes), jams, jellies, pickles, and relishes. Low-acid foods include greens, green beans, root vegetables, peas, peppers, potatoes, and squash. Acid foods are processed in a boiling-water canner. Low-acid foods *must* be processed in a higher-temperature steam-pressure canner.

When canning fruit, it will darken unless you dip it into a commercial antioxidant solution – a mixture of ascorbic and citric acids which should be used according the manufacturer's directions – or a mixture of 1 cup (240 ml) of lemon juice to 2 pints (1 liter) of water.

Fruit should be peeled, cored, or pitted. It can be sliced so that more of it fits in each jar. You will need to use a syrup, juice, or water when canning fruit. A simple syrup can be made from 2½ cups (0.5 kg) of sugar dissolved in 5 cups (1.25 liters) of water. Whichever liquid you prefer, it must be heated and kept hot until you are ready to use it.

Apples
Peel, core, and quarter (or slice) the apples. Boil them gently in syrup for 5 minutes. Fill the hot jars with hot fruit and syrup. Leave ½ in (1.25 cm) headspace. Remove air bubbles, cap, and process them in a boiling-water canner for 20 minutes.

Beans (green or snap)
Use only fresh, tender beans. String, trim, and cut them into 2 in (5 cm) pieces. Pack them tightly into hot jars. Fill the jars with boiling water. Leave 1 in (2.5 cm) headspace. Remove air bubbles, cap, and process them in a steam-pressure canner at 10 lb (5 kg) pressure for 25 minutes.

Berries
Rinse berries in cold water; drain. Measure ½ cup hot syrup into each hot jar. Fill the jars with berries. Leave ½ in (1.25 cm) headspace. Top up with hot water. Remove air bubbles, cap, and process them in a boiling-water canner for 20 minutes.

Carrots
Peel and rinse the carrots. Either slice them or leave them whole, and pack them tightly into hot jars. Fill each jar with boiling water. Leave 1 in (2.5 cm) headspace. Remove air bubbles, cap, and process them in a steam-pressure canner at 10 lb (5 kg) pressure for 30 minutes.

Kale (or any green leafy vegetable such as spinach which becomes dense and compact when cooked.)
Remove any tough stems and wash the greens thoroughly under running water. Chop coarsely into pieces about 1–1½ in (2.5–4cm) across. Place the pieces in a pan with just enough water to cover. Heat, stirring gently, until they wilt. Drain quickly and pack them into hot jars. Fill each jar with boiling water, leaving 1 in (2.5 cm) headspace. Remove air bubbles, cap, and process them in a steam-pressure canner at 10 lb (5 kg) pressure for 70 minutes (pint/500 ml jars) or for 90 minutes (quart/liter jars).

Parsnips and turnips
Peel and slice the vegetables and place them in a pan. Cover them with cold water and bring to a boil for 2 minutes. Drain them quickly and pack them into hot jars. Fill each jar with boiling water. Leave 1 in (2.5 cm) headspace. Remove air bubbles, cap, and process them in a steam-pressure canner at 10 lb (5 kg) pressure for 35 minutes.

Peaches
Remove the skins by dipping the peaches in boiling water and then in cold water. Cut in half, remove pits, and scrape out the dark flesh. Pack hot jars with overlapping slices and fill with hot syrup. Leave ½ in (2.5 cm) headspace. Remove air bubbles, cap, and process them in a boiling-water canner for 30 minutes.

Pears

Peel, core, and quarter the pears. Boil them gently in syrup for 5 minutes and pack the hot fruit into hot jars. Fill each jar with syrup made from 2½ cups (0.5 kg) of sugar dissolved in 5 cups (1.25 liters) of water, leaving ½ in (2.5 cm) headspace. Remove air bubbles, cap, and process them in a boiling-water canner for 25 minutes.

Peas

Shell fresh-picked peas and loosely pack them into hot jars. Fill each jar with boiling water. Leave 1 in (2.5 cm) headspace. Remove air bubbles, cap, and process them in a steam-pressure canner at 10 lb (5 kg) pressure for 40 minutes.

Tomatoes

Remove the skins of the tomatoes by dipping them in boiling water and then in cold water. The skins will slip off. Cut them in half, saving any juice that oozes out. Measure 1 tablespoon of lemon juice into each hot pint (500 ml) jar or 2 tablespoons of lemon juice for a quart (liter) jar. Pack the jars with tomatoes, fill each one with juice or boiling water, and leave ½ in (2.5 cm) headspace. Remove air bubbles, cap, and process them in a boiling-water canner for 90 minutes.

PICKLING

Pickling dates at least from the time of the ancient Greeks and Romans, who preserved perishable foods for winter or times of shortage in this way. There are now whole books filled with recipes for every imaginable kind of pickle, chutney, and relish—once you have mastered the basics, you might want to check your local bookstore or library for more ideas.

Vegetables that are suitable for pickling include broccoli, carrots, cauliflower, small cucumbers, green beans, and zucchini, as well as white pickling onions and green or red peppers.

For the best flavor and appearance, use kosher or canning salt, not regular table salt. Even if you are making a lot of pickles, work in small batches—about 6 lb (3 kg) of vegetables is a manageable quantity and will make 6–7 pints (3–3.5 liters) of pickles.

Pickling in vinegar

1 Peel the vegetables and leave them in heavily salted ice water, for at least 12 hours.

2 Place the vinegar solution (see p.154) in a non-corrosive pan and boil for 5 minutes.

3 Thoroughly rinse and drain the vegetables and add them to the boiling vinegar.

4 Pack the vegetables into hot jars and top up with the vinegar solution.

5 With a plastic spatula, gently pack the vegetables down to dislodge any air bubbles.

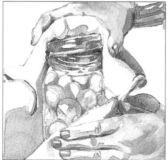

6 Clean the jar and place and center the hot jar lid. Fit and tighten the screw band.

7 When the wire rack is full, lower it into the water and process for 15 minutes.

Prepare your canning jars, lids, and screw bands following the instructions that have already been given (*see page 150*).

Wash and thinly slice 6 lb (3 kg) of vegetables (or, in the case of broccoli and cauliflower, divide them into small florets) and then place them in heavily salted, ice water and leave them there for at least 12 hours, or overnight. If they tend to float, weight them down with a heavy plate. Drain, thoroughly rinse in cold water, and drain again.

In a large pan, combine 3 cups of white distilled vinegar with 2 cups of sugar and 2 tablespoons of salt. Bring to a boil. If you wish, this mixture can be seasoned to your taste with herbs and spices such as garlic, ginger root, mustard seed, peppercorns, rosemary, tarragon, dill, coriander, chili pepper, or oregano. Tie the herbs or spices in cheesecloth, or use a tea-ball, and remove them before filling the jars. Fresh herbs are best; dried herbs will cloud the vinegar.

Add the vegetables to the boiling vinegar, let it return to a boil, then reduce the heat and simmer for 5–15 minutes, depending on the size of the vegetable pieces—just make sure they are thoroughly heated all the way through. Pack the hot vegetables into hot jars, and top up with the vinegar mixture. Leave ¼ in (65 mm) headspace. Remove air bubbles, cap, and process in a boiling-water canner for 15 minutes.

After processing and cooling, leave the pickled vegetables in a cool, dark place for a month or six weeks to cure and develop their full flavor.

JAMS, JELLIES, CONSERVES, AND BUTTERS

Most soft fruits start to deteriorate soon after they are picked. As a result, they have long been preserved in jams, jellies, conserves, and butters. All of these are made by basically boiling fruit or fruit juice with sugar until the mixture sets when cooled.

The pectin content of soft fruit is important when making such preserves, and it largely determines the thickened consistency of the finished produce. Acid is also important since it helps to release the pectin; hence the addition of lemon juice in some recipes. Some fruits contain more pectin than others, and pectin is more abundant in under-ripe rather than overripe fruit. Fruit rich in pectin and acid include cooking apples, sour blackberries, currants, sour plums, concord grapes, and gooseberries. Less pectin is found in blueberries, cherries, peaches, and strawberries. To avoid having to use a lot of sugar, you can add commercial pectin, following the manufacturer's instructions.

Jams can be made with most types of fruit, including apples, pears, plums, greengages, strawberries, apricots, black currants, red currants, blackberries, raspberries, and gooseberries. Black currant and red currant jellies are particularly popular—the latter to accompany game and holiday fare.

Fruit butters and cheeses are made with pureed fruit and these tend to use less sugar than jam. This makes them more attractive healthwise.

Some individual fruit recipes can be found on pages 70–97. Before you start, prepare your canning jars, lids, and screw bands, following the instructions that have already been given.

FREEZING

Most fruits, herbs, and vegetables can be frozen. There are advantages to this method of storage: freezing is quick and easy and it preserves the produce in good condition with its texture and flavor almost intact. All produce should be frozen straight from the garden. Adjust your freezer so that it maintains a temperature of 0° F (-18° C).

Vegetables suitable for freezing include asparagus, beans, broccoli, brussels sprouts, cabbage, carrots, cauli-flower, kale, leeks, parsnips, peas, peppers, spinach, tomatoes, and turnips.

Garden produce can be stored in the freezer in rigid plastic containers or polyethylene freezer bags. Containers are especially useful when the food is fragile, such as asparagus spears, but bags take up less space. Make sure that when you seal the bags that you extract as much air as possible. Always label and date the contents of containers and bags using a permanent marker pen. Freeze the food as rapidly as possible to prevent the formation of ice crystals.

Vegetables should first be prepared by shelling, peeling, or slicing, as required. They must then be blanched in boiling, lightly salted water for 2–5 minutes, drained, plunged into iced water, drained, and thoroughly dried before being put into containers or freezer bags. *Note:* The exceptions are peppers, which should be cut into sections and frozen without blanching, and tomatoes, which can be cored, dipped in boiling water, skinned, and frozen.

When using vegetables from the freezer, the very best way is to defrost them slowly in the refrigerator. If you are in a rush, immerse the frozen food in cold (but never hot) water, or use the defrost cycle on a microwave oven. Greens, such as kale or spinach, which freeze into a solid block, should be defrosted and separated before cooking, or the outside will overcook before the inside is even warm.

Fruit can often be put straight into the freezer, or it may need washing and careful drying first. Some fruits will need topping and tailing, peeling, or having the pits removed before freezing. Fruit suitable for freezing include apples, cherries, blackberries, blueberries, cranberries, currants, gooseberries, pears, plums, raspberries, rhubarb, and strawberries.

Fruit can be frozen in various ways: mixed with sugar (1 part sugar to 4 parts fruit), pureed, dry frozen on trays, or frozen in a sugar syrup (2½ cups [0.5 kg] of sugar dissolved in 5 cups [1.25 liters] of water). Raspberries or blackberries, for example, can be put straight into the freezer on trays; once they are frozen hard, they can be sealed in freezer bags or boxes for storage. Fruit has a tendency to go mushy once it is thawed; defrosting it slowly in the refrigerator, in unopened containers, will help to prevent this.

WINTER STORAGE

It is important that produce picked for natural storage is in perfect condition—anything that is bruised or diseased will cause rotting, which will spread rapidly from one item to another. Look out for onion softness, especially around the necks, or for any black areas on the bulbs. As well, check for maggots in carrots, canker in parsnips causing soft dark patches, and bruising on apples. Do not store any root vegetable that has been damaged by digging or any fruit if the skin has been nicked.

Root vegetables, including carrots, kohlrabi, and turnips, can be stored in deep boxes lined with 1 in (2.5 cm) of slightly damp sand or peat, and they will last well in a cool, dark place. A layer of clean vegetables, with 2 in (5 cm) or so of the leaf stalk left on, should be placed carefully in the box and covered with another layer of peat or sand. Repeat this process until the box is full. The idea is to prevent the produce from drying out and shrivelling, so make sure that the sand or peat is damp but not wet—similar to a wrung-out sponge. Too much moisture will encourage rotting. Place the box in a dry, cool, dark, frost-free place—an unheated cellar is ideal, but a garage may also serve.

Winter squashes can be stored for up to several months on a shelf in a well-ventilated, frost-free area. They can also be hung up in string or netting bags or any other material that allows the free circulation of air around them. Onions

and shallots can be placed on wire or slatted wooden trays and racks, but not too close together since they, also, require the free circulation of air between them. They can also be strung up from a beam or hook by tying the necks with strong twine or raffia. Either way, they should last a whole winter, but inspect them at least weekly and remove any spoiled items.

Parsnips can be left in the ground in some areas and dug up as they are required throughout the winter if you mulch the soil with a thick layer of straw to stop the ground freezing. If space is needed in the garden, parsnips can be dug up and left in a pile where a mild frost will sweeten them and the rain will wash them clean. If several days of hard frost are forecast, cover them with straw and a few old sacks.

Potatoes can be stored in any convenient dark, frost-free place on slatted trays and covered with burlap, paper, or other material to prevent their exposure to light, *which will cause them to turn green and become poisonous*. They can also be placed in boxes lined with straw and covered, layer by layer, with straw or newspaper, or stored in paper or burlap sacks with the top closed. Before being stored, newly dug potatoes should be left to dry off for a couple of days, and they need to be checked fairly regularly for any signs of rotting and the unwanted attention of mice.

Some varieties of apple will keep for several months in cool, dry conditions, especially the old-fashioned russet varieties that have tough skins. They need to be picked carefully in order to avoid bruising them; traditional kitchen gardeners used to pick apples on a clear, dry day and put them carefully into straw-lined baskets to prevent any damage to the fruit. Windfalls or blemished fruit should be cut up, cooked, and frozen, or made into cider. Sound apples can be stored in straw-lined baskets or boxes, in

dry sand or in paper bags; they can be wrapped in tissue paper, placed on wooden trays, or in any container that allows free circulation of air and protects them from sudden changes of temperature. They need to be checked regularly and any rotting fruit removed.

Pears should be picked unripe and will ripen slowly while being stored. They can be placed on shelves, in drawers or shallow boxes, or packed in paper, straw, or dry sand. They need to be kept in a frost-free area and inspected regularly.

DRYING

Many fruit crops, such as apples, apricots, blueberries, cherries, grapes, peaches, pears, plums, strawberries, and tomatoes, can be dried. Grapes, plums, and blueberries should be dipped in boiling water for 30 seconds to remove their natural waxy coating. To prevent darkening, apples, peaches, and pears should be dipped in a commercial antioxidant solution or a mixture of 1 cup (240 ml) of lemon juice with 2 pints (1 liter) water.

In the right climate—sunny and dry with a light breeze and few insects—you can dry fruit outdoors on wooden or plastic racks. Many people prefer the convenience of an electric dehydrator, which, if you like dried fruit and would normally buy a lot of it, can soon pay for itself. Fruit can be rehydrated for use in pies, salads, and so on by soaking it in boiling water for 10 minutes.

The produce needs to be picked very carefully. Avoid any with bruises or signs of mold. Slice produce into even, ½ in (1.25 cm) slices. Once the produce is completely dry, it can be stored in airtight containers in a cool, dark, dry place.

Most herbs can be dried by hanging them in bunches in a well-ventilated place. Once they are bone dry, spread sheets of clean newspaper on the floor and crumble the leaves. Separate the leaves from the twigs and store the herbs in labeled, dated, airtight containers.

To provide health benefits, some of the foodstuffs in this book need simple home-processing into basic medicines. These remedies do not have any unpleasant

preparing home remedies

side-effects, and many of the herbs, such as peppermint and chamomile, make refreshing, delicious beverages in their own right. When following the more complicated recipes on preserving, it is vital to follow the instructions for sterilizing all bottles, jars, and equipment.

Natural remedies

Natural remedies are rewarding to make at home and can be prepared for both internal and external use. They range from the simple application of fruit, vegetable, or herb leaf to the skin as for an insect bite; to the addition of the ingredient to a meal—mint with lamb, for example, and the preparation of an infusion, decoction, or tincture for internal use or as the basis for an ointment or salve. Fresh herbs, fruits, and vegetables picked straight from the garden, or those which you have grown, harvested, and preserved, can be used to make home preparations that are not only far cheaper than store-bought products, but also provide the additional satisfaction of being home-made and containing only top quality natural ingredients which have been grown organically.

INTERNAL USE

Taken internally, in whatever form, the therapeutic constituents of plants enter the bloodstream via the digestive tract. The two most popular forms of internal remedies are infusions and decoctions.

Infusions

Infusions, which can either be taken as remedies or as relaxing or revitalizing teas, are made from the leaves and flowers of either fresh or dried plants. Traditionally, fragrant herbs such as chamomile, sage, lemon balm, and peppermint have been used in this way, but many leaves, like Indian or China tea, can be prepared with boiling water. However, plants which contain a high percentage of mucilage such as borage should be made up with cold water and left to infuse for 10–12 hours, as the hot water may destroy their therapeutic constituents. It is best to make fresh infusions each day, but if they are stored in airtight containers, homemade infusions can keep for up to 2 days in a refrigerator. Herbal infusions can also be used as mouthwashes, gargles, douches, and hair rinses, as well as in foot baths and on compresses.

There are no hard and fast rules about temperature, although infusions and decoctions are usually taken hot for fevers, congestion, colds, and skin problems; they are taken lukewarm or cold for problems associated with the kidneys and urinary tract such as cystitis and urethritis, and can be enjoyed as cooling drinks in the summer. If you find that certain herbs are too bitter, try combining the herbs required for the remedy with more palatable ones, such as lemon balm, peppermint, or lavender, or you can use honey, unrefined sugar, or licorice as a sweetener.

TO MAKE AN INFUSION

2 oz (50 g) fresh plant or
1 oz (25 g) dried leaves or
   flowers
1 pint (570 ml) boiling water

Dosage: A normal adult dosage, for both infusions and decoctions (opposite) is 1 cup, taken 3 times a day in the case of chronic problems, and 6 or more times a day in the case of acute illness.

1 Warm the jug or pot with hot water and add the plant.

2 Pour the boiling water over the plant. Cover the container and leave to infuse for 5–10 minutes.

3 Strain and drink, adding honey or sugar to taste, or cover and store in a refrigerator for use within 2 days.

Decoctions

Decoctions are similar to infusions but involve a boiling process to break down the harder woody parts of a plant—the stalks, seeds, and roots—before their therapeutic constituents can be absorbed by the water. Some herbs can be bought ready-powdered, or they can be broken up by pounding with a pestle and mortar (if the roots are fresh) or ground up in a coffee grinder. Decoctions, which taste stronger than infusions, can be drunk as teas or incorporated into syrups, gargles, compresses, and douches.

TO MAKE A DECOCTION

2 oz (50 g) fresh plant or
1 oz (25 g) dried leaves
22 fl oz (650 ml) cold water

Dosage: Some remedies require double-strength infusions and decoctions. For these, simply double the amount of plant used.

1 Place plant in a stainless steel or enameled pan (not aluminum) and cover with water.

2 Bring to a boil, cover, and simmer for 10–20 minutes.

3 Take off the heat, strain, and either store in a sterilized jar in a refrigerator for up to 2 days or drink immediately.

Tinctures

Tinctures are taken in small doses: they can be taken internally, added to bath water, or used in ointments and lotions. The fresh or dried herb is softened by being soaked in a mixture of alcohol and water, which should be measured according to the ratios laid down in an herbal pharmacopeia. This ratio varies from plant to plant: for many herbs, the liquid comprises 25% alcohol and 75% water, but for some resinous herbs the amount of alcohol needs to be increased in order to extract the therapeutic constituents often to 45% alcohol and sometimes even 90%. A standard recipe is 1 part fresh herb to 2 parts fluid or 1 part dried herb to 5 parts fluid. Brandy or vodka, which are 45% alcohol, provide a good alcohol solution, and both can be used in the recipe below. For children (and adults) who require a sweeter taste, you can prepare syrup-like tinctures using equal parts of water and glycerol. For more watery fresh plants, such as lemon balm and peppermint, use 80% glycerol.

TO MAKE A TINCTURE

7 oz (200 g) dried or
1 lb (500 g) fresh plant
1 3/4 pints (1 liter) 45% alcohol solution or glycerol solution (*see above*)

Dosage: Normal dosage is 1 teaspoon for adults or 10 drops–1/2 teaspoon for children, 3 times a day with or after meals. Take every 2 hours in acute illness. Dilute with water.

1 Place plants in a jar with an airtight lid and cover with alcohol or glycerol solution. Seal and store in a refrigerator for 2 weeks, shaking daily.

2 Strain the mixture into a jar or jug through a cheese-cloth bag.

3 Pour into a dark, labeled bottle and store in a refrigerator.

Syrups

Syrups incorporating infusions, decoctions, or tinctures can be used to make natural remedies more palatable to both children and squeamish adults.

The standard dosage is 2 teaspoons for children, 3–4 times a day for chronic problems and 6–8 times a day in acute illness. If you have no infusion, decoction, or tincture prepared, 1 teaspoon of herbs or finely chopped fruit or vegetables can be mixed with honey and given to children (1 teaspoon 3 times a day for chronic problems and 1 teaspoon every 2 hours in acute illness).

TO MAKE A SYRUP USING AN INFUSION OR DECOCTION

12 oz (350 g) 1:1 mixture of thin honey and unrefined sugar
1/2 pint (300 ml) double strength infusion or decoction of herbs, fruits, or vegetables

1 Heat infusion or decoction with honey/sugar mixture in a stainless steel or enameled pan.

2 Stir mixture as it starts to thicken and skim off foam from surface.

3 Let cool before pouring into a cork-topped bottle. Use required dose or store in a refrigerator for up to 3 days.

TO MAKE A SYRUP USING A TINCTURE

1/2 pint (300 ml) boiling water
13 fl oz (360 ml) thin honey or 1 lb 5 oz (600 g) unrefined sugar
4 fl oz (120 ml) tincture of your choice

1 Pour boiling water over honey or sugar and stir over a low heat until the sugar dissolves and the water starts to boil.

2 Remove from the heat and add tincture in a ratio of 1 part tincture to 3 parts syrup. If placed in an airtight, sterilized container, this syrup can be stored indefinitely.

Herbal and Fruit Vinegars

Some tinctures can be made with undiluted cider or wine vinegar producing herbal vinegars such as garlic or rosemary vinegar for culinary use, bay or sage vinegar for use as skin and hair preparations, or raspberry vinegar for treating children's coughs and sore throats.

TO MAKE HERBAL OR FRUIT VINEGAR

Enough fresh herbs or fruit to fill your chosen container loosely
Enough cider or wine vinegar to cover the herbs

1 Bruise the freshly picked herbs or fruit and place them in the sterilized container.

2 *Pour on the vinegar, seal with an airtight lid, and leave for 2 weeks in a sunny position, shaking daily. If at the end of this time a stronger taste is required, strain the vinegar and repeat with fresh herbs or fruit.*

3 *For culinary uses, either store as it is or strain and re-bottle. If it is for use as a hair rinse add 1 fl oz (30 ml) vinegar to 8 fl oz (220 ml) of water just before use.*

EXTERNAL USE

There are several external pathways by which plant constituents can be introduced into the bloodstream: through the skin, through the rectum or vagina, through the nasal passages by means of inhalation, and through the conjunctiva of the eye. Remedies can take the form of gargles and mouthwashes, inhalants, and eyewashes, hair rinses and herbal baths, suppositories, salves, creams, poultices, and compresses. Infusions, decoctions, and tinctures are all used in these remedies, as well as infused and essential oils.

Gargles and mouthwashes

These can be made from infusions, decoctions, or from tinctures: dilute 1 teaspoon (5 ml) of tincture in 2 fl oz (50 ml) of water, or use half a cup of infusion or decoction. Gargle 2–3 times a day for chronic problems, and every 2 hours if the infection is acute. If you are using the mixture as a mouthwash, use 2–3 times daily.

Poultices and compresses

Both poultices and compresses are applied to areas of pain and swelling. The difference between them is that while poultices use the plant or foodstuff itself, compresses use an extract of it. Bread poultices can be used for bringing boils to a head, and cabbage-leaf poultices are good for relieving the pain of arthritic joints and for mastitis. Plants such as cabbage leaves can be applied directly to the skin. For painful joints, simply, remove any hard ribs and outer leaves, warm the leaves in hot water or by placing them over a radiator, apply to the affected part and bandage.

Compresses can also be applied to painful joints, and they are useful for soothing skin irritations. Cold compresses are sometimes used to help relieve headaches and migraines, and to bring down fevers.

TO MAKE A WARM POULTICE

Plant, fresh or dried
2 pieces of gauze
A light cotton bandage

1 *If using fresh herbs, bruise them using a pestle and mortar, or soak in hot water for several minutes to soften them. If using dried herbs, add hot water to make a paste.*

2 *Place sufficient plant to cover the affected area between two pieces of gauze.*

3 *Use a light cotton bandage to bind the poultice to the affected area. Keep it warm with a hot water bottle. Leave for 2–4 hours or overnight, depending on the severity of the problem.*

TO MAKE A COMPRESS

Hot or cold infusion or
  decoction of your choice or
  a few drops of essential oil
  in water
Face cloth or small towel

*1 Soak face cloth or towel in a
bowl of the chosen
preparation, and wring out
excess liquid.*

*2 Apply to affected area,
repeating several times.*

Herbal baths

Herbal baths can provide relaxation,
clear blocked noses, and soothe aching
limbs. Infusions and decoctions may be
strained and poured into the water, a
cheesecloth bag filled with fresh or
dried aromatic herbs may be hung
beneath the hot tap, or a few drops of
infused or essential oils may be added to
the bath water. Warm water has the
effect of opening the pores of the skin,
so that when herbal preparation are
added to a bath, the plant constituents
are quickly absorbed via the skin and in
the case of volatile oils, via inhalation
through the nasal passages into the
lungs and thus into the bloodstream.
Add 1 pint (570 ml) double-strength
infusion or decoction and soak in the
bath for 15 to 30 minutes for best effect.
You can also add essential oils to your
bath. These, because they are extracted
from plants by steam distillation, cannot
be made at home, but they can be
bought from health-food stores and other
suppliers of natural products. Note that
they should always be diluted in a base
oil for babies, young children, and
people with sensitive skins (2 drops of
essential oil per 1 tsp [5 ml] of base oil).

Hand and foot baths

This is an excellent way to give
remedies to babies and children; it is an
easy way of introducing plant
constituents into the bloodstream,
since both the hands and the feet are
sensitive areas with plenty of nerve
endings. Use either 1 qt (1 L) of
double-strength infusion or decoction,
several teaspoons of tincture, or a few
drops of dilute essential oil in a bowl of
hot water (warm for babies and
children). Foot baths should be taken in
the evening (for 8 minutes; 4 minutes
for children), and hand baths in the
morning, for the same length of time.

Salves and creams

Salves can be made by macerating
plants in oil. Creams can be made by
stirring tinctures, infusions, decoctions,
or a few drops of essential oil into
aqueous (water-based) cream. At the
pharmacist's, look for a skin cream that
has water as the first item on the
ingredients label.

TO MAKE A SALVE

¾ pint (425 ml) olive oil
2 oz (55 g) beeswax
Fresh or dried herb

*1 Mix oil and beeswax
together in a heatproof bowl.*

*2 Add as much of your chosen
herb as the mixture will cover,
and mix together.*

*3 Heat gently over a saucepan
of boiling water for 2–3 hours.*

4 *Press out through a cheese-cloth bag, discard the herb, and pour the warm oil into a sterilized jar. Allow to solidify.*

5 *Use or store in a cool place. It will keep for up to 3 months.*

TO MAKE A CREAM

2–3 drops essential oil
2 oz (55 g) aqueous cream

Mix together thoroughly and smooth into the skin. Made with chamomile oil, this recipe is especially good for eczema.

Infused oils

Both dilute essential oils and infused oils can be used as massage oils for inflamed and painful joints, or for neuralgia, bruises, and swellings. They can also be used as bath oils, and as a base for ointments or liniments. There are two methods of preparing these, and both can be easily done at home (*see below*). Not for internal use.

TO MAKE A COLD INFUSED OIL

Cold-pressed safflower, walnut, or almond oil to fill chosen jar
Enough fresh or dried plant to pack chosen sterilized jar

1 *Pack jar with your chosen plant and cover with oil. Seal with an airtight lid. Leave in a sunny place for 2 weeks, shaking daily.*

2 *Squeeze the oil through a cheesecloth bag into a jug.*

3 *Pour strained oil into dark-colored, airtight, sterilized storage bottles and label.*

TO MAKE A HOT INFUSED OIL

1 ¾ pint (1 liter) cold-pressed safflower, walnut, or almond oil
Enough fresh or dried plant, chopped, to fill the top part of a double boiler

1 *Place the chopped plant in the top part of a double boiler and cover with oil.*

2 *Place the double boiler on the heat and simmer for 2–3 hours.*

3 *Allow to cool before straining into dark-colored, airtight, sterilized storage bottles.*

Treatment Chart

	Abrasions	Acid indigestion	Acne	Alcoholism	Allergies	Anemia	Anxiety/tension	Appetite, poor	Arteriosclerosis	Arthritis	Asthma	Atherosclerosis	Babies' colic	Babies' sleeping problems	Back pain	Bladder infections	Bleeding gums	Boils & abscesses	Bowel disorders/infections	Bronchial congestion	Bronchitis	Bruises & sprains	Burns & scalds, min...
Apple										•													
Apricot						•	•	•										•					
Artichoke			•					•	•	•		•											
Arugula						•															•	•	
Asparagus										•													
Basil	•						•																
Beans, dried																							
Beans, green						•				•													
Blackberry						•		•				•					•		•			•	
Black currant								•				•											
Blueberry/bilberry	•											•					•		•				
Borage					•		•			•	•												
Brassicas	•			•		•				•								•				•	
Calendula																		•					
Caraway								•															
Carrot	•					•				•								•		•		•	
Celery										•													
Chamomile							•			•			•					•				•	•
Cherry						•				•													
Chervil								•		•													
Chicory										•													
Chives						•						•						•					
Coriander				•				•		•													
Cranberry								•															
Cucumber										•						•							
Dandelion								•										•					
Dill								•					•	•									
Fennel								•		•											•		
Garlic													•										
Gooseberry																		•					
Horseradish								•		•								•		•			
Lavender	•						•	•		•								•					•
Leek	•											•						•	•				•
Lemon balm	•				•		•	•															
Lettuce							•																
Marjoram							•											•					
Marrow																		•					•
Mint													•	•									
Onion						•				•								•			•		
Parsley						•	•			•					•							•	
Parsnip																							
Pea																							
Peach		•					•											•					
Pear		•								•													
Pepper, hot																							
Pepper, sweet					•																		
Plum						•				•								•					
Potato	•	•						•		•													•
Pumpkin																		•					•
Radish								•												•			
Raspberry						•	•										•						
Red currant								•		•		•											
Rhubarb																							
Rosemary							•	•															•
Sage	•																					•	•
Sorrel			•			•												•					
Spinach						•		•															
Squash																		•					•
Strawberry			•																				
Sweet bay										•													
Thyme							•			•	•						•	•					
Tomato						•																	
Turnip			•							•							•	•					
Zucchini																		•					•

	Candidiasis	Cardiovascular problems	Cataracts	Chest infections	Chilblains	Childhood infections	Circulatory disease	Colds	Cold sores	Colic	Colitis	Color & night vision problems	Congestion	Conjunctivitis	Constipation	Coughs	Cramp	Croup	Cuts	Cystitis	Depression, mild	Diarrhea	Digestive problems	Diverticulitis
Apple								●					●		●	●			●		●			
Apricot															●									
Artichoke																								
Arugula								●					●	●										
Asparagus			●												●									
Basil								●		●			●		●				●		●			
Beans, dried													●		●	●			●					●
Beans, green															●									
Blackberry								●					●		●						●			
Black currant				●				●					●		●						●			
Blueberry/bilberry													●						●		●			
Borage						●		●					●								●			
Brassicas				●				●			●		●		●									
Calendula	●			●				●			●								●					
Caraway								●		●			●		●									
Carrot												●	●						●	●				
Celery															●					●				
Chamomile											●			●	●				●				●	
Cherry													●		●	●					●			
Chervil																								
Chicory															●									
Chives																								
Coriander			●					●		●					●	●						●		
Cranberry											●				●					●				
Cucumber																								
Dandelion															●									
Dill										●			●		●	●						●	●	
Fennel										●					●									
Garlic		●		●		●		●					●		●									
Gooseberry															●									
Horseradish				●				●					●		●	●								
Lavender																			●					
Leek				●				●					●		●				●		●			
Lemon balm						●													●					
Lettuce								●					●		●	●								
Marjoram		●	●					●		●			●				●							
Marrow											●													
Mint		●						●	●				●			●					●			
Onion								●					●		●					●				
Parsley										●														
Parsnip															●									
Pea															●									
Peach															●				●			●		
Pear										●					●				●			●		
Pepper, hot				●				●							●									
Pepper, sweet		●																						
Plum															●									
Potato				●							●		●						●					●
Pumpkin											●													
Radish								●					●			●								
Raspberry								●					●		●	●						●		
Red currant				●				●					●		●	●						●		
Rhubarb															●							●	●	
Rosemary				●				●					●		●									
Sage										●			●					●	●					
Sorrel																								
Spinach															●									
Squash											●													
Strawberry						●									●									
Sweet bay				●				●		●			●			●						●		
Thyme				●				●					●			●	●			●		●		
Tomato															●									
Turnip				●				●					●		●	●								
Zucchini											●													

Treatment Chart

	Dysentery	Earache	Eczema	Fevers	Flatulence	Flu	Fluid retention	Gall bladder problems	Gastritis	Gastroenteritis	Gingivitis	Gout	Hangovers	Hay fever	Headaches	Heart & arterial disease	Heartburn	Heat rash	Heavy periods	Hemorrhoids	Hiccups	High blood pressure	High cholesterol
Apple				•			•	•	•						•								
Apricot																							
Artichoke			•				•					•					•						
Arugula																							
Asparagus							•																
Basil					•										•								
Beans, dried																			•			•	
Beans, green							•					•											
Blackberry	•			•		•			•									•	•				•
Black currant				•		•	•					•											
Blueberry/bilberry				•															•				
Borage				•		•	•																
Brassicas									•			•					•						
Calendula				•		•	•	•												•			
Caraway					•																•		
Carrot			•				•					•		•									
Celery												•										•	•
Chamomile		•	•					•															
Cherry							•					•											
Chervil							•					•							•				
Chicory							•	•				•			•		•						
Chives																							
Coriander				•	•	•			•														
Cranberry																							
Cucumber			•	•			•											•					
Dandelion							•	•															
Dill					•																		
Fennel					•	•											•						
Garlic																						•	•
Gooseberry																							
Horseradish				•		•	•					•		•									
Lavender					•										•								
Leek					•																		
Lemon balm			•											•									
Lettuce									•														
Marjoram				•		•									•								
Marrow									•						•								
Mint				•							•				•								
Onion					•	•	•					•										•	•
Parsley					•		•					•			•								
Parsnip																							
Pea																							•
Peach							•		•								•						
Pear							•		•			•					•						
Pepper, hot				•						•													
Pepper, sweet																							
Plum					•		•					•											
Potato									•												•		
Pumpkin									•						•								
Radish			•				•	•				•											
Raspberry				•		•	•																•
Red currant				•		•	•					•											
Rhubarb																							
Rosemary				•		•	•						•		•								
Sage					•		•				•	•											
Sorrel			•				•																
Spinach							•																•
Squash									•						•								
Strawberry				•			•											•					
Sweet bay				•	•	•	•					•			•								
Thyme				•		•				•	•												
Tomato							•										•						
Turnip			•				•					•											
Zucchini									•						•								

	Hormonal problems	Hot flushes	Hyperacidity	Hyperactivity	Indigestion	Infections	Inflammatory eye problems	Inflammatory problems	Insect bites & stings	Insomnia	Intestinal infections	Irritable bowel syndrome	Joint pain	Kidney stones	Lethargy	Liver problems	Low immunity	Mastitis	Measles	Menopausal problems	Migraines	Morning sickness	Mouth ulcers	Muscle tension
Apple			●	●							●				●									
Apricot															●									
Artichoke				●																				
Arugula															●									
Asparagus															●									
Basil				●				●							●					●				●
Beans, dried																								
Beans, green																								
Blackberry																						●		
Black currant						●		●											●			●		
Blueberry/bilberry							●															●		
Borage	●															●	●							
Brassicas				●											●	●	●							
Calendula		●														●								
Caraway				●											●									
Carrot											●					●								
Celery																								
Chamomile				●						●										●				
Cherry															●									
Chervil															●									
Chicory				●											●	●								
Chives						●											●							
Coriander				●											●									
Cranberry													●				●							
Cucumber						●		●																
Dandelion													●	●	●		●							
Dill				●																				●
Fennel				●															●					
Garlic				●											●									
Gooseberry																●								
Horseradish				●											●									
Lavender				●					●	●										●				
Leek								●	●															
Lemon balm				●		●			●															
Lettuce										●		●												
Marjoram										●														●
Marrow				●																				
Mint												●			●					●			●	
Onion															●									
Parsley								●							●									
Parsnip																								
Pea															●									
Peach				●											●									
Pear												●			●									
Pepper, hot																								
Pepper, sweet																								
Plum															●									
Potato																								
Pumpkin				●																				
Radish																●								
Raspberry				●																	●	●		
Red currant					●			●											●		●			
Rhubarb					●																			
Rosemary				●		●									●					●				
Sage		●		●																			●	
Sorrel																								
Spinach															●		●							
Squash				●																				
Strawberry					●	●																		
Sweet bay				●											●									
Thyme																	●						●	
Tomato								●																
Turnip																	●							
Zucchini				●																				

Treatment Chart

	Muscular aches & pains	Nausea	Nerve pain	Nervous indigestion	Nervous palpitations	Neuralgia	Night sweats	Overheating	Pain in childbirth	Peptic ulcers	Period pains	Pharyngitis	PMS	Poor circulation	Poor concentration	Poor vision	Poor lactation	Prostate problems	Respiratory infections	Rheumatism	Rhinitis	Ringworm	Sinusitis
Apple										●									●				
Apricot			●																				
Artichoke		●																					
Arugula														●									
Asparagus																							
Basil				●																			●
Beans, dried																							
Beans, green																							
Blackberry																							
Black currant																							●
Blueberry/bilberry																							
Borage															●								
Brassicas															●								●
Calendula									●	●													
Caraway														●									
Carrot																●		●					
Celery																							
Chamomile		●				●							●										
Cherry																							
Chervil														●									
Chicory																							
Chives														●									
Coriander								●															●
Cranberry																							
Cucumber								●															
Dandelion																		●					
Dill		●																					
Fennel		●									●						●						
Garlic														●									
Gooseberry																							
Horseradish														●									●
Lavender	●	●			●									●						●			●
Leek														●									
Lemon balm	●	●								●		●									●		
Lettuce				●				●															
Marjoram										●				●									
Marrow							●		●								●						
Mint		●				●				●												●	●
Onion												●								●			●
Parsley										●				●									
Parsnip																							
Pea																							
Peach																							
Pear																							
Pepper, hot						●				●				●									●
Pepper, sweet																		●					
Plum																							
Potato																							
Pumpkin							●	●										●					
Radish																			●	●			●
Raspberry		●							●														
Red currant																							●
Rhubarb																							
Rosemary														●									●
Sage		●					●			●			●						●				
Sorrel																							
Spinach																							
Squash							●	●									●						
Strawberry																							
Sweet bay	●	●									●			●						●			
Thyme														●									●
Tomato																							
Turnip																							
Zucchini							●	●									●						

	Skin problems/infections	Sluggish digestion	Sore throats	Stomach infections	Stomach ulcers	Stress/stress-related problems	Stress-related digestive problems	Sunburn	Thrush	Tiredness	Tonsillitis	Ulcers (open sores)	Urethritis	Urinary infections	Urticaria	Vaginal infections	Varicose veins	Viruses	Vitamin/mineral deficiency	Vomiting	Warts	Weak digestion	Worms	Wounds
Apple	●																							●
Apricot										●									●		●			
Artichoke														●										
Arugula										●											●			
Asparagus	●									●				●					●				●	
Basil			●							●														
Beans, dried																								
Beans, green				●						●									●					
Blackberry	●				●									●		●	●							
Black currant	●		●											●										
Blueberry/bilberry	●		●						●					●										
Borage						●				●														
Brassicas	●																●		●					
Calendula	●								●			●									●			●
Caraway			●							●														
Carrot																			●					
Celery				●										●							●			
Chamomile			●	●		●			●						●									
Cherry										●				●					●					
Chervil		●								●														
Chicory										●				●										
Chives		●		●						●														
Coriander	●	●								●														
Cranberry														●										
Cucumber								●							●									
Dandelion	●									●				●					●					
Dill		●					●																	
Fennel																								
Garlic										●													●	
Gooseberry																								
Horseradish	●									●														
Lavender			●			●																		
Leek														●										
Lemon balm																		●						
Lettuce																								
Marjoram					●									●										
Marrow																							●	
Mint	●		●							●	●													
Onion																							●	
Parsley										●									●					
Parsnip										●									●					
Pea										●														
Peach	●					●				●		●							●					
Pear						●				●									●					
Pepper, hot																								
Pepper, sweet																								
Plum	●									●									●					
Potato					●			●					●											
Pumpkin																							●	
Radish											●											●		
Raspberry			●								●													
Red currant	●		●																					
Rhubarb																								
Rosemary										●														
Sage			●								●			●		●					●			
Sorrel	●																		●					
Spinach	●									●									●		●			
Squash																							●	
Strawberry	●						●												●					
Sweet bay										●														
Thyme			●						●		●													
Tomato																			●					
Turnip																			●					
Zucchini																							●	

GLOSSARY

Amino acid One of the building blocks of protein.

Analgesic Substance that relieves pain.

Apiol Constituent of volatile oil in celery.

Arteriosclerosis Hardening of the arteries.

Antioxidant Prevents cell damage by free radicals.

Atherosclerosis Build-up of plaque on the inner lining of the arteries.

Betacarotene Precursor of vitamin A.

Bioflavonoids Plant chemicals that enhance the action of vitamin C and have an antioxidant action.

Carcinogen Substance associated with tumor formation, such as cancer.

Chlorophyll Green coloring in leaves that traps the energy of sunlight for photosynthesis to occur.

Cholecystitis Inflammation of the gall bladder.

Cholesterol Type of lipid and the most abundant steroid in the body. In excess, can cause arterial disease.

Coxsackie Virus that occurs in the intestinal tract.

Earthing up To build up earth around the base of a plant such as fennel.

Enzyme Protein molecules that act as catalysts for chemical reactions in the body.

Folic acid One of the B vitamins.

Free radicals Molecules formed by oxygen metabolism that can damage the body's cells.

Hemoglobin Pigment found in red blood cells that carries oxygen.

Hypertension High blood pressure.

Indole Nitrogen compounds thought to protect against

cancer by enhancing the elimination of estrogen.

Mucilage Gel-like substance from certain plants used as a laxative.

Nephritis Inflammation of the kidneys.

Oxalic acid Substance that inhibits the absorption of some minerals and can predispose to kidney stones.

Pectin Soluble fiber that helps to regulate the bowels and lower the level of blood cholesterol.

Peristalsis Muscular movements of the gut, enabling the passage of food and fluid.

Pharyngitis Inflammation of the pharynx (throat).

Phenolic acid Naturally occurring substance with antiseptic and disinfectant properties.

Phytochemicals Plant-derived chemicals.

Phyto-estrogens Substances found in plants

that have effects similar to estrogen.

Protease Enzyme that digests proteins.

Purines Compounds, often found in high-protein foods, that form uric acid when metabolized.

Rhizome Creeping underground stem of some plants.

Sulfites Sulfur compounds used in food preservatives that can trigger asthma attacks.

Tannins Yellow or brown compounds derived from plants and used in medical astringents to prevent infections or reduce inflammation.

Triglycerides Fats found in the body, high levels of which may indicate coronary artery disease.

Vertical Garden Space-saving garden with upward-growing plants.

Volatile oils Compounds with an antiseptic action in highly scented herbs.

USEFUL ADDRESSES (U.S.A.)

NURSERIES AND GARDENS

Brooklyn Botanic Garden
1000 Washington Avenue
Brooklyn, NY 11225-1099
Tel. (718) 622-4433

Caprilands Herb Garden
534 Silver Street
Coventry, CT 06238
Tel. (860) 742-7244

DeBaggio Herbs
923 N Ivy Street
Arlington, VA 22201
Tel. (703) 243-2498
(Note: no mail order)

Gilbertie's Herb Gardens
65 Adams Road
PO Box 118
Easton, CT 06612
Tel. (203) 452-0913

Logee's Greenhouses
141 North Street
Danielson, CT 06239
Tel. (860) 774-8038

Nichols Garden Nursery
1190 North Pacific Hwy NE
Albany, OR 97321
Tel. (541) 928-9280

Peconic River Herb Farm
2749 River Road
Calverton, NY 11933
Tel. (516) 369-0058

United States National Arboretum
3501 New York Avenue NE
Washington, DC 20002-1958
Tel. (202) 245-2726

SEED SOURCES

The Cook's Garden
PO Box 535
Londonderry, VT 05148
Tel. (802) 824-3400

Johnny's Selected Seeds
310 Foss Hill Road
Albion, ME 04910
Tel. (207) 437-9294

Seeds Blum
Idaho City Stage
Boise, ID 83706
Tel. (208) 342-0858

Seeds of Change
PO Box 15700
Santa Fe, NM 87506-5700
Tel. (800) 957-3337

Shepherd's Garden Seeds
30 Irene Street
Torrington, CT 06790
Tel. (860) 482-3638

Southern Exposure Seed Exchange
PO Box 170
Earlysville, VA 22936
Tel. (804) 973-4703

GENERAL

The Herb Society of America
9019 Kirtland Chardon Road
Kirtland, OH 44094
Tel. (216) 256-0514

USEFUL ADDRESSES (CANADA)

GARDENS

Devonian Botanic Garden
University of Alberta
Edmonton, AB T6G 2E1
Tel (403) 987-3054

Montreal Botanical Garden
(Jardin botanique de Montréal)
4101 Sherbrooke Street East
Montreal, QC H1X 2B2
Tel (514) 872-1400

Royal Botanical Gardens
680 Plains Road West
Burlington, ON L7T 4H4
Tel (905) 527-1158

SEEDS, PLANTS, PRODUCTS

Island Seed Co.
P.O. Box 4278
Depot 3
Victoria, BC V8X 3X8
Tel (250) 744-3677

Aubin Nurseries Ltd.
P.O. Box 1089
Carman, MB R0G 0J0
Tel (204) 745-6703

Gardenimport Inc.
Box 760
Thornhill, ON L3T 4A5
Tel (905) 731-1950

William Dam Seeds
P.O. Box 8400, Dundas, ON L9H 6M1
Tel (905) 628-6641

W. H. Perron & Co. Ltd.
2900 Blvd. Labelle
Laval, QC H7P 5FB
Tel (514) 682-9768

GENERAL

Canadian Botanical Association
c/o Dr. C.C. Chinnappa
Dept. of Biological Sciences
University of Calgary
Calgary, AL T2N 1N4
Tel (403) 220-7465

INDEX

Hyphenated page numbers following ailments indicate that there is scattered (not continuous) reference to these within the given pages.

PRINTED IN BELGIUM BY
proost
INTERNATIONAL BOOK PRODUCTION